# Liquid Dielectrics in an Inhomogeneous Pulsed Electric Field

# IOP Plasma Physics Series

## Series Editors


**Richard Dendy**
*Culham Centre for Fusion Energy and the University of Warwick, UK*

**Uwe Czarnetzki**
*Ruhr-University Bochum, Germany*


### About the Series

The IOP Plasma Physics ebook series aims at comprehensive coverage of the physics and applications of natural and laboratory plasmas, across all temperature regimes. Books in the series range from graduate and upper-level undergraduate textbooks, research monographs and reviews.

The conceptual areas of plasma physics addressed in the series include:

- Equilibrium, stability and control
- Waves: fundamental properties, emission, and absorption
- Nonlinear phenomena and turbulence
- Transport theory and phenomenology
- Laser-plasma interactions
- Non-thermal and suprathermal particle populations
- Beams and non-neutral plasmas
- High energy density physics
- Plasma-solid interactions, dusty, complex and non-ideal plasmas
- Diagnostic measurements and techniques for data analysis

The fields of application include:

- Nuclear fusion through magnetic and inertial confinement
- Solar-terrestrial and astrophysical plasma environments and phenomena
- Advanced radiation sources
- Materials processing and functionalisation
- Propulsion, combustion and bulk materials management
- Interaction of plasma with living matter and liquids
- Biological, medical and environmental systems
- Low temperature plasmas, glow discharges and vacuum arcs
- Plasma chemistry and reaction mechanisms
- Plasma production by novel means

# Liquid Dielectrics in an Inhomogeneous Pulsed Electric Field

**M N Shneider**
*Princeton University, Princeton, NJ, USA*

**M Pekker**
*George Washington University, Washington, DC, USA*

**IOP** Publishing, Bristol, UK

ISBN 978-0-7503-1245-5 (ebook)
ISBN 978-0-7503-1246-2 (print)
ISBN 978-0-7503-1247-9 (mobi)

DOI 10.1088/978-0-7503-1245-5

Version: 20160801

IOP Expanding Physics
ISSN 2053-2563 (online)
ISSN 2054-7315 (print)

British Library Cataloguing-in-Publication Data: A catalogue record for this book is available from the British Library.

Published by IOP Publishing, wholly owned by The Institute of Physics, London

IOP Publishing, Temple Circus, Temple Way, Bristol, BS1 6HG, UK

US Office: IOP Publishing, Inc., 190 North Independence Mall West, Suite 601, Philadelphia, PA 19106, USA

*To our parents*

# Contents

# Preface

This book comprehensively describes the phenomena that occur in liquid dielectrics under the influence of inhomogeneous pulsed electric fields. It is based on experimental and theoretical studies performed over the last few years, in many of which the authors themselves were directly involved. This book includes data about the material properties of polar and non-polar liquid dielectrics. It presents the dynamics of dielectric fluids under the action of ponderomotive forces. The book reviews the conditions for the formation of cavitation, its dynamics, and its observation by optical methods in a variety of liquid dielectrics (from water to superfluid helium). It also discusses the cavitation mechanism of nanosecond and sub-nanosecond breakdown in liquid dielectrics. This book is intended for a broad audience, from students to engineers and scientists, who are interested in the current research questions in the electrodynamics and hydrodynamics of liquid dielectrics.

Scientific studies of the behavior of dielectric fluids in an electric field began with the invention of the capacitor in 1745 by Cunaeus (and independently by Musschenbroek), earning it the name Cunaeus Leyden jar, in honor of the university where he worked. The term dielectric permittivity was introduced by Faraday in 1837. A well-known example of the effect of an electric field on a dielectric fluid is drawing it into a charged capacitor. A rigorous mathematical description of this effect can be found in many textbooks (e.g. [1–3]).

During the last hundred years, the investigation of the effect of electric fields on dielectric fluids has mainly focused on the study of dielectric breakdown strength at high electric fields and the dependence of the dielectric permittivity of liquid dielectrics on the frequency of electromagnetic waves [4–12]. Some interest in dielectric fluids in inhomogeneous pulsed electric fields originated from an area of physics that is not related to fluid dynamics at all. Experiments from [13] showed that when a nanosecond pulse voltage is applied to a needle-like electrode in water, discharge occurs at electric fields that are much lower than expected from estimates of the electric strength of water based on its density. The mystery was that the breakdown in the liquid arose without the formation of gas bubbles, in which electrons would be able to gain enough energy to ionize water molecules. In other words, the normal breakdown mechanism associated with ohmic heating of water followed by bubble formation in the high-voltage area did not take place [14–16]. Bubbles would not have been able to form during the time of an order of 1–5 ns and they were not observed in the experiment.

A bubble-free mechanism of breakdown in liquid dielectrics was proposed in [17], and further developed and experimentally studied by many authors [18–25]. As already noted, in an inhomogeneous electric field, the dielectric liquid tends to flow toward the greater electric field area. This motion gives rise to the gradient of hydrostatic pressure that compensates the tension in the fluid associated with the action of the ponderomotive forces (a macroscopic volumetric force acting on the fluid caused by electromagnetic fields) [1, 2]. In [26], we showed that the velocity of the induced flow in the fluid near a pointed electrode, where the gradient of the electric field is maximal, does not exceed a few meters per second. Hence, fluid motion, and therefore changes in hydrostatic pressure, are negligible during pulses of the order of 0.1–3 nanoseconds, which allows a negative pressure to arise in the electrode vicinity that reaches an absolute value of tens or even hundreds of megapascals. It is well known in hydrodynamics that discontinuities may appear in negative pressure areas in liquids, as described by the standard theory of cavitation (e.g. [27, 28]). These discontinuities just serve as 'bubbles', in which electrons gain sufficient energy for ionization and, hence, for the initiation of sub-nanosecond or nanosecond breakdown in liquid. We also showed in [28] how rapidly the voltage at the electrode must increase so that the negative pressure produced by electrostrictive forces is sufficient for the appearance of microscopic cavitation discontinuities (nanopores) in the liquid.

The main topic of this book is related to the cavitation effects in dielectric fluids caused by the electrostrictive ponderomotive forces in inhomogeneous pulsed electric fields. In [29, 30] we considered the conditions for the growth of cavitation nanopores in a liquid dielectric and in [29] we showed that three characteristic regions can be distinguished in the vicinity of the electrode. In the first region, where the electric field gradient is greatest, the cavitation nanopores that form during the voltage nanosecond pulse may grow to the size at which an electron accelerated by the field inside the pores can acquire enough energy to excite and ionize the liquid on the opposite pore wall (i.e. the breakdown conditions are satisfied). In the second region, the negative pressure caused by electrostriction is large enough for initiating cavitation (which can be registered by optical methods) but, during the voltage pulse, the pores do not reach the size at which the potential difference across their borders becomes sufficient for ionization or excitation of water molecules. In the third region the development of cavitation is impossible, due to an insufficient level of negative pressure: the spontaneously occurring micropores do not grow and collapse under the influence of surface tension forces.

## References

[1] Tamm I E 2003 *Fundamentals of the Theory of Electricity* (Moscow: Fizmatlit)

[2] Landau L D and Lifshitz E M 1991 *Electrodynamics of Continuous Media* (*A Course of Theoretical Physics* vol 8) 3rd edn (Oxford: Pergamon)

[3] Panofsky W K H and Phillips M 1962 *Classical Electricity and Magnetism* (Reading, MA: Addison-Wesley)

[4] Koch M, Fischer M and Tenbohlen S 2007 The breakdown voltage of insulation oil under the influence of humidity, acidity, particles and pressure *Int. Conf. APTADM* (*Wroclaw, Poland, 26–28 September* 2007)

[5] Nagel M and Leibfried T 2006 Investigation on the high frequency, high voltage insulation properties of mineral transformer-oil *IEEE Conf. Electrical Insulation and Dielectric Phenomena (15–18 October 2006)*

[6] Hasted J B 1972 Liquid water: dielectric properties *The Physics and Chemistry of Water* vol 1 ed F Franks (New York: Plenum) chapter 7

[7] Fernandez D P, Mulev Y, Goodwin A H R and Levelt Senders J M H 1995 A database for the static dielectric constant of water and steam *J. Phys. Chem. Ref. Data* **24** 33

[8] Kaatze U 1989 Complex permittivity of water as a function of frequency and temperature *J. Chem. Eng. Data* **34** 371

[9] Gadani D H, Rana V A, Bhatnagar S P, Prajapati A N and Vyas A D 2012 Effect of salinity on the dielectric properties of water *Indian J. Pure Appl. Phys.* **50** 405

[10] Lizhi H, Toyoda K and Ihara I 2008 Dielectric properties of edible oils and fatty acids as a function of frequency, temperature, moisture and composition *J. Food Eng.* **88** 151

[11] Buchner R, Barthel J and Stauber J 1999 The dielectric relaxation of water between 0 °C and 35 °C *Chem. Phys. Lett.* **306** 57

[12] Yeh I-C and Berkowitz M L 1999 Dielectric constant of water at high electric fields: molecular dynamics study *J. Chem. Phys.* **110** 7935

[13] Starikovskiy A, Yang Y, Cho Y I and Fridman A 2011 Nonequilibrium plasma in liquid water: dynamics of generation and quenching *Plasma Sources Sci. Technol.* **20** 024003

[14] Ceccato P 2009 Filamentary plasma discharge inside water: initiation and propagation of a plasma in a dense medium *PhD Thesis* LPP École Polytechnique, Palaiseau

[15] Ushakov V Y, Klimkin V F and Korobeynikov S M 2005 *Breakdown in Liquids at Impulse Voltage* (Tomsk: NTL) (in Russian)
Ushakov V Y, Klimkin V F and Korobeynikov S M 2007 *Impulse Breakdown of Liquids (Power Systems)* (Berlin: Springer) (Engl. transl.)

[16] An W, Baumung K and Bluhm H 2007 Underwater streamer propagation analyzed from detailed measurements of pressure release *J. Appl. Phys.* **101** 053302

[17] Shneider M N, Pekker M and Fridman A A 2012 Theoretical study of the initial stage of sub-nanosecond pulsed breakdown in liquid dielectrics *IEEE Trans. Dielectr. Electr. Insul.* **19** 1597

[18] Dobrynin D, Seepersad Y, Pekker M, Shneider M N, Friedman G and Fridman A 2013 Non-equilibrium nanosecond-pulsed plasma generation in the liquid phase (water, PDMS) without bubbles: fast imaging, spectroscopy and leader-type model *J. Phys. D: Appl. Phys.* **46** 105201

[19] Seepersad Y, Pekker M, Shneider M N, Dobrynin D and Fridman A 2013 On the electrostrictive mechanism of nanosecond-pulsed breakdown in liquid phase *J. Phys. D: Appl. Phys.* **46** 162001

[20] Seepersad Y, Pekker M, Shneider M N, Dobrynin D and Fridman A 2013 Investigation of positive and negative modes of nanosecond pulsed discharge in water and electrostriction model of initiation *J. Phys. D: Appl. Phys.* **46** 355201

[21] Pekker M, Seepersad Y, Shneider M N, Fridman A and Dobrynin D 2014 Initiation stage of nanosecond breakdown in liquid *J. Phys. D: Appl. Phys.* **47** 025502

[22] Seepersad Y, Fridman A and Dobrynin D 2015 Anode initiated impulse breakdown in water: the dependence on pulse rise time for nanosecond and sub-nanosecond pulses and initiation mechanism based on electrostriction *J. Phys. D: Appl. Phys.* **48** 424012

[23] Starikovskiy A 2013 Pulsed nanosecond discharge development in liquids with various dielectric permittivity constants *Plasma Sources Sci. Technol.* **22** 012001

[24] Marinov I L, Guaitella O, Rousseau A and Starikovskaia S M 2013 Cavitation in the vicinity of the high-voltage electrode as a key step of nanosecond breakdown in liquids *Plasma Sources Sci. Technol.* **22** 042001

[25] Marinov I, Starikovskaia S and Rousseau A 2014 Dynamics of plasma evolution in a nanosecond underwater discharge *J. Phys. D: Appl. Phys.* **47** 224017

[26] Shneider M N and Pekker M 2013 Dielectric fluid in inhomogeneous pulsed electric field *Phys. Rev. E: Sci. Instrum.* **87** 043004
[27] Brennen C E 2013 *Cavitation and Bubble Dynamics* (Cambridge: Cambridge University Press)
[28] Kim K-H, Chahine G, Franc J-P and Karimi A 2014 *Advanced Experimental and Numerical Techniques for Cavitation Erosion Prediction* (Berlin: Springer)
[29] Shneider M N and Pekker M 2015 Pre-breakdown cavitation development in the dielectric fluid in the inhomogeneous, pulsed electric fields *J. Appl. Phys.* **117** 224902
[30] Pekker M and Shneider M N 2015 Pre-breakdown cavitation nanopores in the dielectric fluid in the inhomogeneous, pulsed electric fields *J. Phys. D: Appl. Phys.* **48** 4240009

# Acknowledgement

We are grateful to our colleagues and friends for stimulating and encouraging discussions, in particular A V Starikovskiy, S M Starikovskaia, V Semak, I Marinov, D Dobrynin, and Y Seepersad. We are thankful to M Solway, who provided substantial assistance in editing the book. We are also grateful to Lora Shneider for her understanding and support of our work.

# Author biographies

## Mikhail N Shneider

Dr Mikhail N Shneider received a master's degree in theoretical physics (with honors) from the Kazan State University, Russia, a PhD in plasma physics and chemistry from the All-Union Electrotechnical Institute, Moscow, and Doctor of Sciences (the highest scientific degree in Russia) in plasma physics and chemistry from the Institute for High Temperatures, Russian Academy of Sciences, Moscow. Since 1998, Dr Shneider has worked at the Mechanical and Aerospace Engineering Department, Princeton University. At present he is a Senior Scientist in the Applied Physics Group. His research interests are the theoretical study of gas discharge physics, physical gas dynamics, biophysics, atmospheric electrical phenomena, non-linear optics, and laser–matter interaction. Dr Shneider has been invited many times as a guest professor to universities in China, France, Germany, Great Britain, and Russia. He has published more than 170 papers in refereed journals (including eight review papers), three US patents and one book.

## Mikhail Pekker

Dr Mikhail Pekker received a master's degree in physics and applied mathematics from the Novosibirsk State University, Russia, and PhD in physics and mathematics from the Institute of Theoretical and Applied Mechanics, Novosibirsk, Russia. From 1993–2007 he worked at the Institute for Fusion Studies of the University of Texas at Austin, from 2010–14 at the Drexel Plasma Institute of Drexel University as an Assistant Research Professor, and since 2015 he has worked as a research scientist in the Department of Mechanical and Aerospace Engineering at George Washington University. His research interests are the theoretical study of gas discharge physics and biophysics. Dr Pekker has published about 100 papers, including two review papers.

# Nomenclature

| | |
|---|---|
| $\vec{D}$ | electric displacement |
| $\vec{E}$ | electric field |
| $F_r$, $F_\theta$ | components of the ponderomotive electrostrictive forces near the pore |
| $I(r, \theta)$ | intensity of the scattered radiation |
| $I_I$ | intensity of the incident radiation |
| $I_i$ | ionization potential |
| $I_n$ | affinity energy of an electron in a negative ion |
| $N_b$ | number of cavitation nanopores |
| $N_e$ | number of electrons |
| $P = P_{in} - P_{out}$ | pressure difference at the boundary of the bubble |
| $P_{cr}$ | critical negative pressure at which cavitation begins |
| $P_E$ | pressure associated with the electrostrictive ponderomotive forces |
| $P_L$ | Laplace pressure |
| $\vec{P}_d$ | dipole moment |
| $P_-$ | negative pressure |
| $R_b$ | radius of nanopore |
| $R_{0,b}$ | initial radius of nanopore |
| $R_{cr}$ | radius of nanopore corresponding to the critical pressure $P_{cr}$ |
| $T$ | temperature |
| $U$ | voltage at an electrode |
| $U_0$ | maximum voltage at an electrode |
| $V_{cr}$ | volume of nanopore of the critical radius |
| $W_b$ | energy required to create a bubble (nanovoid) of radius $R_b$ |
| $W_{cr}$ | energy required to create a bubble (nanovoid) of radius $R_{cr}$ |
| $W_{diel}$ | electrostatic energy |
| $Y_b(r, \theta)$ | scattering factor |
| $Y_{\Omega,b}(r)$ | scattering factor integrating over the solid angle |
| $c$ | speed of light |
| $c_s$ | speed of sound in liquid |
| $e$ | electron charge |
| $e_i$ | intrinsic internal energy per unit mass |
| $\vec{f}_p$ | volumetric density of the ponderomotive force |
| $\hbar$ | Planck's constant |
| $n$ | refractive index |
| $n$ | density of molecules |
| $n_b$ | density of cavitation nanopores |
| $n_d$ | density of polar molecules |
| $k_\sigma$ | Tolman factor |
| $p$ | hydrostatic pressure |
| $p_0$ | undisturbed hydrostatic pressure |
| $\vec{p}_d$ | density of the dipole moment |
| $p_{d,n}$ | projection of the density of the dipole moment of on the normal to the surface |
| $p_{total} = p + P_E$ | total pressure |
| $q$ | electric charge |
| $r_{el}$ | radius of curvature of the needle electrode |
| $s$ | entropy per unit mass |

| | |
|---|---|
| $\vec{u}$ | fluid velocity |
| $\vec{u}_{\rm n}$, $\vec{u}_{\rm s}$ | velocities of normal and superfluid liquids |
| $w(I_{\rm n}, E)$ | probability of electron detachment |
| $\Gamma$ | rate of nanopores generation per unit time per unit volume |
| $\Delta\phi$ | potential difference between the nanopore's poles |
| $\Sigma$ | electrical conductivity |
| $\alpha$ | effective polarizability of the spherical cavity in dielectric media |
| $\alpha_{\rm E} = \frac{\partial \varepsilon}{\partial \rho}\rho$ | |
| $\alpha_{\rm e}$ | electronic polarizability of atoms and molecules |
| $\alpha_{\rm orient}$ | orientational polarizability for polar molecules |
| $\alpha_{\rm polar}$ | effective polarizability of polar molecules |
| $\beta_{\rm E} = \frac{\varepsilon_0}{2}\frac{\partial \varepsilon}{\partial \rho}\rho$ | |
| $\delta_{\rm b}$ | size of the transition region at the liquid–gas interface boundary |
| $\varepsilon$ | relative dielectric permittivity of the medium |
| $\varepsilon_0$ | vacuum permittivity |
| $\eta$ | dynamic viscosity |
| $\kappa$ | attenuation coefficient |
| $\lambda$ | laser wavelength in a vacuum |
| $\vec{\mu}_{\rm d}$ | dipole moment of the polar molecule |
| $\mu_{\rm d,0}$ | absolute value of the dipole moment of the polar molecule |
| $\nu$ | kinematic viscosity |
| $\nu$ | frequency of electromagnetic wave |
| $\rho$ | density of a liquid |
| $\rho_0$ | unperturbed fluid density |
| $\rho_{\rm f}$ | volume density of free charge |
| $\rho_{\rm bound}$ | volume density of the bound charges |
| $\rho_{\rm n}$, $\rho_{\rm s}$ | densities of normal and superfluid liquids |
| $\sigma_{\rm bound}$ | surface density of bound charge |
| $\sigma_{\rm f}$ | surface density of free charges |
| $\sigma_{\rm s}$ | surface tension coefficient |
| $\sigma_{\rm 0s}$ | surface tension coefficient for a planar interface boundary |
| $\tau_0$ | rise time of the voltage pulse |
| $\tau_{\rm M}$ | Maxwell time |
| $\tau_{\rm d}$ | relaxation time for dipoles in a polar dielectric |
| $\phi$ | potential |
| $\chi$ | dielectric susceptibility |
| $\omega$ | angular frequency of the laser radiation |

**M N Shneider and M Pekker**

# Chapter 1

# Introductory description of processes related to negative pressure in liquids

*In this chapter we briefly consider cavitation, negative pressure, the Rayleigh model of bubble dynamics, the Zel'dovich-Fischer theory of nucleation, electrostrictive phenomena in liquid dielectrics, and give estimates for cavitation inception in a non-uniform electric field.*

## 1.1 A qualitative picture of the formation of discontinuities in a liquid

Without going into the history of the science of cavitation, note the following fact. Engineers and shipbuilders were first faced with cavitation in 1894 at marine trials of the British destroyer Daring. Its huge propeller screws should have given the ship a speed of 32 knots (59 km $h^{-1}$), which was unprecedented at the time. However, the speed reached during the sea trials was much slower. The engineers changed the screw sizes and profiles, but the result was still the same; the rate was much lower than expected. Most importantly, in the process of testing, they observed unexplained erosion of the rotor blades, as if some invisible force had knocked out bits of high-strength steel from the surface (figure 1.1). Further studies showed that the observed erosion was associated with the occurrence of bubbles formed on the surface of the blades. They observed that bubbles did not appear at low speeds of screw rotation and, consequently, the associated erosion was absent. The problem was solved when three propellers with substantially reduced speed were installed to replace the sole propeller. The phenomenon of the emergence of bubbles on the surface of rapidly rotating propellers became known as cavitation.

Some understanding of the reasons behind cavitation appeared only in 1917, when Rayleigh published a paper in which he considered the problem of expansion and contraction of gas bubbles in a liquid located in the regions of stretching

doi:10.1088/978-0-7503-1245-5ch1  

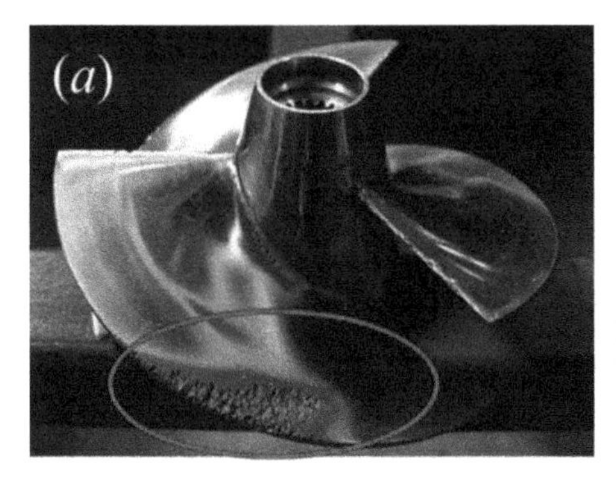


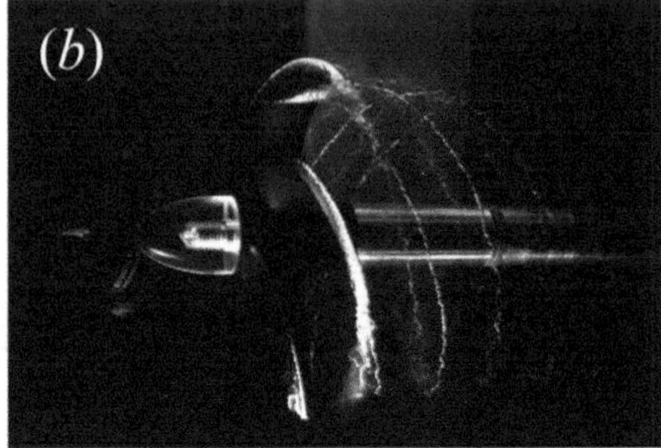


**Figure 1.1.** The effect of cavitation on screw-propellers. (*a*) shows the erosion zone of the propeller due to cavitation and (*b*) shows the formation of cavitation bubbles in the areas of negative pressure. This cavitation image has been obtained by the author(s) from the Brittanica website [1] where it was made available by Erik Axdahl under a CC BY-SA 2.5 license. It is included within this book on that basis. It is attributed to Erik Axdahl. Image (*b*) is courtesy of the Cavitation Research Laboratory, University of Tasmania and the Defence Science and Technology Group, Australia.

negative pressure (directed outward from the bubble) [2]. According to Rayleigh, the role of negative pressure was played by the saturated vapor pressure inside the bubble. However, this work did not answer the main question: how did the bubbles appear in the liquid if they were not there initially. Speculations about impurities, such as dust particles or dissolved gases that supposedly may serve as centers for the formation of bubbles, cannot withstand criticism, because bubbles form on rotating screws in distilled water the same way as they do in 'dirty' water. The present understanding of the nature of cavitation came 25 years after Rayleigh's work, with the appearance of the theory of nucleation, proposed by Zel'dovich [3] and later independently by Fisher [4]. In these papers, they showed that thermal fluctuations in the liquid may lead to the appearance of nanobubbles with sizes sufficient for the stretching negative pressure to exceed the pressure caused by surface tension (Laplace pressure), which tends to reduce the size of the bubble.

## 1.2 Negative pressure

Here, we quickly summarize the concept of negative pressure in liquids, which is not widely covered in common textbooks. Indeed, the pressure in practical reality is always positive, for example in ideal gas physics, while it is only the pressure difference that can be negative. In contrast, 'negative pressure' is a key concept in the fluid mechanics of liquids with stretching tension that violates homogeneity and thereby leads to cavitation. We will discuss this effect in detail in subsequent chapters. Interestingly, modern cosmology explains the expansion of the Universe using negative pressure, which is described by a cosmological constant $\Lambda$ [5].

Within the frame of high-school physics, a pressure in a liquid is a scalar function, which is independent of the forces it is caused by (the Pascal law). This is why the initial radius $R$ of a rubber ball dipped into a liquid to a depth of $h \gg R$, where it experiences isotropic pressure $P_{\text{out}} \approx \rho g h$ (figure 1.2(a)), will reduce to a value $R_1 < R$.

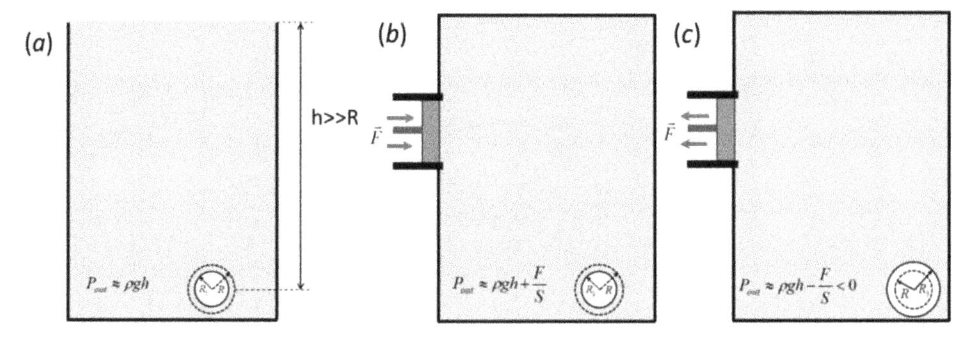


**Figure 1.2.** A rubber ball of radius $R$ dipped into a vessel filled with water. In (a) the vessel is open, so compression by the water column reduces the radius of the ball to $R_1 < R$. In (b) the same vessel with the same amount of water is sealed and a piston exerts a force onto the water, which compresses the ball further to a radius $R_2 < R_1$. In (c) the piston is switched to decompress the water with a large enough force such that the total pressure on the ball is negative and its radius increases to $R_3 > R$.

Next, we consider a sealed vessel with water and a piston pushing with force $\vec{F}$ against the water (figure 1.2(b)). In this case, a ball experiencing the isotropic pressure $P_{\text{out}} \approx \rho g h + |F|/S$ (where $S$ is the surface of the piston), will reduce in radius to $R_2 < R_1$. Finally, if $\vec{F}$ is switched in the opposite direction (figure 1.2(c)), then the surface of the ball will experience the constant isotropic pressure $P_{\text{out}} \approx \rho g h - |F|/S$. Consequently, $P_{\text{out}} < 0$ if $|F|/S > \rho g h$, such that the ball will be isotropically stretched instead of being shrunk. In this respect, a positive pressure does not conceptually differ from the negative one, unless the latter forms discontinuities. Due to isotropy of the stretching tension, such a discontinuity acquires a spherical shape and it can either expand in a Rayleigh-like manner, due to the saturated vapor pressure (if the stretching negative pressure exceeds the Laplace one), or collapse due to the mitigation or suppression of the stretching tension.

Electrostrictive forces acting on a liquid dielectric in a non-uniform electric field also produce a negative pressure in the dielectric, which is isotropic in a homogenous single-phase liquid.

## 1.3 The Rayleigh bubble [2]

In the spherically symmetric case, the continuity equation for an incompressible fluid outside of the bubble is (figure 1.3):

$$\vec{\nabla} \cdot \vec{u} = \frac{1}{r^2}\frac{\partial(r^2 u)}{\partial r} = 0, \tag{1.1}$$

where $u \equiv u_r$ is the velocity of the fluid. This approximation is valid if the rate of expansion (compression) of the bubble is less than the speed of sound. Equation (1.1) implies that

$$r^2 u = R_b^2 U_b, \tag{1.2}$$

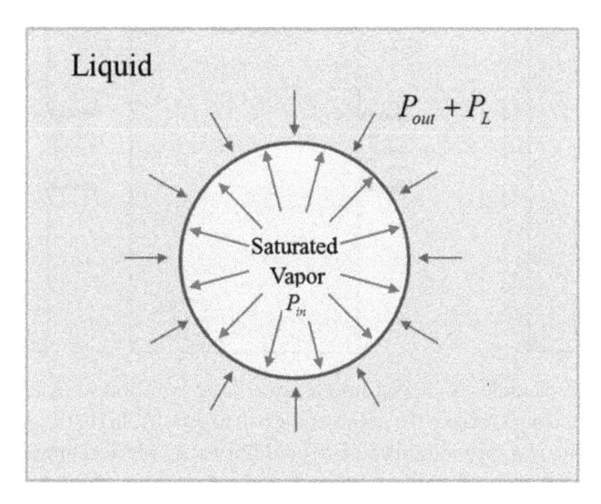


**Figure 1.3.** A vapor bubble in a liquid. $P_{in}$ and $P_{out}$ are the vapor pressure and the pressure in the liquid at the bubble–liquid interface. In addition to pressure $P_{out}$, there is also the Laplace pressure $P_L = 2\sigma_s/R_b$, caused by the surface tension, acting on the surface of the bubble.

where $R_b$ is the radius of the bubble and $U_b = dR_b/dt$ is the velocity of the boundary of the bubble. Then

$$u = \frac{R_b^2}{r^2} U_b, \quad r \geqslant R_b. \tag{1.3}$$

Accordingly, the kinetic energy of a fluid with a density $\rho_0$ at a time $t$ is

$$W_k = \frac{1}{2} \int_{r>R(t)}^{\infty} 4\pi\rho_0 u^2 r^2 dr = 2\pi\rho_0 R^3 \left(\frac{dR_b}{dt}\right)^2. \tag{1.4}$$

The work of pressure to displace fluid from the area of the bubble is equal to

$$A_b = \int_{t_0}^{t} 4\pi \left(P_{in}(t') - P_{out}(t') - \frac{2\sigma_s}{R_b(t')}\right) R_b^2(t') \frac{dR_b(t')}{dt'} dt'. \tag{1.5}$$

Here, $R_{0,b}$ is the initial radius of the bubble, $P_{out}$ is the hydrostatic pressure exerted by the fluid on the boundary of the bubble, $P_{in}(t')$ is the value of the vapor pressure in the bubble, $\frac{2\sigma_s}{R_b(t')}$ is the pressure associated with the surface tension of the liquid (Laplace pressure), and $\sigma_s$ is the surface tension coefficient. By equating the work of the pressure forces (1.5) to the kinetic energy of water (1.4), we obtain

$$2\pi\rho_0 R_b^3 \left(\frac{dR_b}{dt}\right)^2 = \int_{t_0}^{t} 4\pi \left(P_{in}(t') - P_{out}(t') - \frac{2\sigma_s}{R_b(t')}\right) R_b(t') \frac{dR_b}{dt'} dt' \tag{1.6}$$

or

$$\frac{\rho_0}{2} \frac{d}{dt} \left(R_b^3 \left(\frac{dR_b}{dt}\right)^2\right) = \left(P_{in} - P_{out} - \frac{2\sigma_s}{R_b}\right) R_b \frac{dR_b}{dt}. \tag{1.7}$$

Equation (1.7) describes the dynamics of a spherically symmetric bubble in the case where any fluid motion is only associated with the expansion (compression) of the bubble. In other words, on the order of a few radii of the bubble, the external pressure $P_{out}$ can be considered constant. In the case of constant $P_{in}$ and $P_{out}$, the bubble expands at

$$R_b > \frac{2\sigma_s}{P_{in} - P_{out}}, \tag{1.8}$$

and compresses at

$$R_b < \frac{2\sigma_s}{P_{in} - P_{out}}. \tag{1.9}$$

## 1.4 Zel'dovich–Fisher nucleation

The energy $W_b$ required to create a bubble (nanovoid) of radius $R_b$ is equal to the work (1.5) with the opposite sign. In the case of constant pressures $P_{in}$ and $P_{out}$,

$$W_b = 4\pi\sigma_s R_b^2 - \frac{4}{3}\pi R_b^3 (P_{in} - P_{out}). \tag{1.10}$$

Figure 1.4 shows the dependence of $W_b$ on the radius. The maximum of $W_b$ is achieved when

$$R_{cr} = \frac{2\sigma_s}{P_{in} - P_{out}}, \tag{1.11}$$

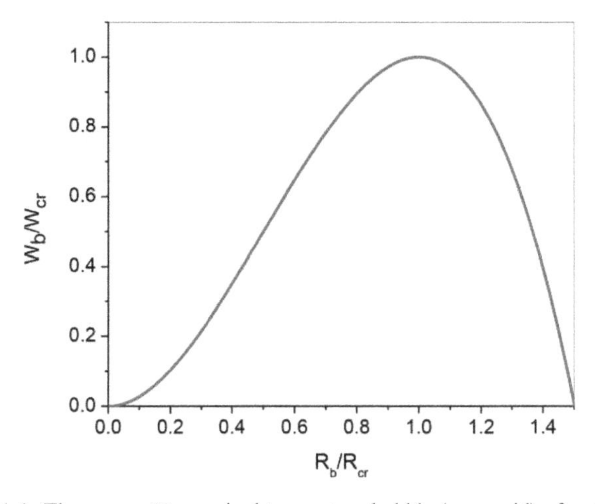


**Figure 1.4.** The energy $W_b$ required to create a bubble (nanovoid) of radius $R_b$.

and is equal to

$$W_{cr} = W_b(R_{cr}) = \frac{16\pi\sigma_s^3}{3(P_{in} - P_{out})^2}. \tag{1.12}$$

Since the fluctuation of energy $W_b(R_b)$ corresponds to the fluctuation of the radius $R_b$, the probability that a bubble of radius $R_{cr}$ will form is proportional to $e^{-W_{cr}/k_BT}$, where $k_B$ is the Boltzmann constant and $T$ is the temperature of the fluid.

The probability of the creation of the critical bubble in the volume $V$ during time $t$ due to the presence of thermal fluctuations, in accordance with the theory of [3, 4, 6–9] is

$$W_{pore} = 1 - \exp\left(-\int_0^t \int_V \Gamma \mathrm{d}t_1 \mathrm{d}V\right), \tag{1.13}$$

where $V$ and $t$ are the volume and the duration of the measurement, respectively. Here, $\Gamma[\mathrm{m}^{-3}\,\mathrm{s}^{-1}]$ characterizes the rate of creation of the cavitation voids per unit volume per second. The integration in (1.13) is carried out in time and over the entire area of observation.

In general, the expression for the nucleation rate is as follows:

$$\frac{\mathrm{d}n_b}{\mathrm{d}t} = \Gamma = \frac{1}{V_{cr}}\frac{1}{\tau_{exp}} = \Gamma_0 \exp\left(-\frac{W_{cr}}{k_BT}\right)[\mathrm{m}^{-3}\mathrm{s}^{-1}], \tag{1.14}$$

where $V_{cr} = 4\pi R_{cr}^3/3$ is the critical volume and $\tau_{exp}$ is the expectation time for appearance of the pore with the critical radius $R_{cr}$; $\Gamma_0$ is the kinetic prefactor, which depends on the theoretical model used. For example, in [6, 9], they used

$$\Gamma_0 \approx \frac{1}{V_{cr}}\frac{1}{\Delta t_T} = \frac{3}{4\pi R_{cr}^3}\frac{k_BT}{2\pi\hbar}, \tag{1.15}$$

where $\frac{1}{\Delta t_T} = \frac{k_BT}{2\pi\hbar}$ is the effective thermal frequency and $\hbar$ is the Planck constant. The corresponding inverse expectation time is $\frac{1}{\tau_{exp}} = \frac{1}{\Delta t_T}\exp(-\frac{W_{cr}}{k_BT})$.

It is possible to use a more detailed expression for the kinetic prefactor, for instance as in [4],

$$\Gamma_0 = N\frac{k_BT}{2\pi\hbar}\exp\left(-\frac{\Delta f^*}{k_BT}\right), \tag{1.16}$$

where $N$ is the density of the molecules in liquid and $\Delta f^*$ is the free energy of activation for the motion of an individual molecule of liquid past its neighbors into or away from a bubble surface. By multiplying the numerator and denominator of the right-hand side of equation (1.16) by $V_{cr}$, we arrive at the expression

$$\Gamma_0 = \frac{3}{4\pi R_{cr}^3}\frac{k_BT}{2\pi\hbar}n_{cr}\exp\left(-\frac{\Delta f^*}{k_BT}\right), \tag{1.17}$$

where $n_{cr} = NV_{cr}$ is the number of molecules in the pore volume. Since the critical radius of the pores (as will be shown later) is about 1–3 nm, then $n_{cr} \sim 10^2 - 3 \times 10^3$. Considering that $\Delta f^*/k_B \sim 1000 - 2500$ K (2–5 kcal mol$^{-1}$) [4], we obtain $n_{cr} \exp(-\frac{\Delta f^*}{k_B T}) \sim 1 - 10$, and hence formula (1.17) is reduced to (1.15). Note that there are alternative expressions for $\Gamma_0$, where, instead of the effective thermal frequency $\frac{k_B T}{2\pi\hbar}$, parameters obtained from other physical considerations are used. For example, based on theory [3, 7, 8]

$$\Gamma_0 = N\left(\frac{2\sigma_s}{\pi MB}\right)^{1/2}, \tag{1.18}$$

where $B$ is a factor that depends on the pressure: $2/3 < B < 1$; $M$ is the mass of the fluid molecules; and $\sigma_s$ is the surface tension coefficient. However, for example, in the case of water, the values of $\Gamma_0$, as estimated by formulas (1.15) and (1.18), differ by less than one order of magnitude. In general, as noted in [6], an exact knowledge of $\Gamma_0$ is not so important for estimating the critical value of the negative pressure due to the exponential dependence of the nucleation rate on $W_{cr}/k_B T$.

Based on (1.14), the number of fluctuations of the critical radius occurring per unit volume per unit time [3, 4] is

$$\frac{dn_b}{dt} = \Gamma = \frac{1}{4\pi R_{cr}^3/3}\frac{k_B T}{2\pi\hbar}\exp\left(-\frac{W_{cr}}{k_B T}\right) = \frac{3P^3}{16\pi\sigma_s^3}\frac{k_B T}{2\pi\hbar}\exp\left(-\frac{16\pi\sigma_s^3}{3k_B T \cdot P^2}\right). \tag{1.19}$$

Here, $P = P_{in} - P_{out}$ is the pressure difference at the boundary of the bubble.

The critical pressure at which cavitation occurs can be easily estimated by equating the exponential factor in (1.19) to 1:

$$|P_{cr}| \sim \sqrt{16\pi\sigma_s^3/3k_B T}. \tag{1.20}$$

Taking into account that $\sigma_s = 0.072$ Nm$^{-2}$ for water at the temperature $T \approx 300$ K [10], the corresponding $P_{cr} \approx -1200$ MPa. Table 1.1 shows measurements of the critical pressure for cavitation initiation in water obtained by different methods. These results are significantly less than the estimate (1.20), because this estimate is obtained for the coefficient of surface tension corresponding to a plane interface between the liquid and the vacuum. In fact, the size of the transition layer $\delta_b$ on the liquid–gas (liquid–vacuum) boundary is finite and is comparable to the radius of the critical bubble (void) $R_{cr}$. Therefore, for a small nanovoid, the surface tension of the liquid cannot be considered as a size-independent constant. Introducing a correction factor $k_\sigma$,

$$\sigma_s = k_\sigma\, \sigma_{0s}, \tag{1.21}$$

(where $\sigma_{0s}$ is the surface tension coefficient for the plane surface) we rewrite (1.19) as [9]

$$\frac{dn_b}{dt} = \Gamma = \frac{3k_B T}{16\pi(k_\sigma\sigma_{0s})^3}\frac{|P|^3}{4\pi\hbar}\exp\left(-\frac{16\pi(k_\sigma\sigma_{0s})^3}{3k_B T \cdot P^2}\right). \tag{1.22}$$

**Table 1.1.** The threshold values of negative pressure for the initiation of cavitation in water measured by various methods.

| Method | $P_{cr}$ [MPa] | Ref. |
|---|---|---|
| Berthelot | −16 | [11] |
| Berthelot | −18.5 | [12] |
| Centrifugation | −27.7 ($T = 10$ °C) | [13] |
| | −2 ($T = 0$ °C) | |
| | −22 ($T = 50$ °C) | |
| Shock wave | −27 | [14] |
| Acoustic | −21 | [15] |
| Inclusions | −140 | [16] |
| Acoustic | −24 | [17] |
| Acoustic | −1,−2 | [18] |

If, in accordance with [16], we assume that the negative pressure at which cavitation begins is $P_{cr} \approx -30$ kPa, then $k_\sigma \approx 0.25$ follows from (1.14) and accordingly the critical radius of the pore is $R_{cr} = 2k_\sigma \, \sigma_{0s}/|P_{cr}| \approx 1$ nm.

The obtained radius of a critical pore is not sufficiently precise, because the coefficient $k_\sigma$ should depend on the size of the pore. When the pore radius is of the order of $R \approx 100$ nm, $k_\sigma$ should be close to 1, and $k_\sigma$ vanishes as $R$ tends to zero. In practice, it is convenient to use the phenomenological formula for the surface tension correction as a function of the radius of the bubble proposed by Tolman in [19]:

$$k_\sigma = 1/(1 + 2\delta_b/R_{cr}). \tag{1.23}$$

However, it is rather problematic to verify the accuracy of this approximation for $k_\sigma$ at $\delta_b \sim R_{cr}$, because it is very difficult to detect individual nanopores in a liquid.

The experimental determination of the critical negative pressure is a very complicated problem, because the sizes of the 'prospective' emerging nanopores are on the order of a few nanometers and the characteristic volumes of their appearance are on the order of several tens of cubic microns. The large spread of values presented in table 1.1 may be explained by these factors, as well as by different degrees of water purity. We will show in chapter 8 that laser Rayleigh scattering allows one to study the development of cavitation in volumes of a few cubic microns when the radii of incipient nanopores are on the order of 1–10 nm [20].

## 1.5 A qualitative description of the processes in a liquid dielectric in a non-uniform pulsed electric field

When a non-uniform electric field is turned on, volumetric electrostrictive (ponderomotive) forces arise almost instantly and tend to displace the dielectric fluid into the region of maximum field. It is natural to expect that if the tension in the dielectric exceeds a certain threshold, then ruptures will form. Ponderomotive forces, in addition to creating a tension in the volume, drive fluid into a motion resulting in increased density in the region of the maximum electric field near the electrode. Since the density and pressure are related by the equation of state (see chapter 3), the growth of density in the region of maximum field leads to an increase in hydrostatic pressure and its gradient. After some time, this begins to compensate for the ponderomotive force. The rise time of the field is crucial. If the field is switched on so slowly that the fluid has time to flow to the electrode, reducing the tension to below the threshold, then a discontinuity cannot develop. If, on the other hand, the field increases so rapidly that the fluid does not have time to come into motion due to inertia, then the electrostrictive tension may reach the threshold for fracture, as in a solid dielectric, and cavitation may develop.

In chapter 5, we will show that a nanosecond voltage pulse applied to a needle-like electrode is sufficient for the emergence of discontinuities (nanopores, ruptures, bubbles) in the liquid. In chapter 6, we will show that due to the discontinuity of the dielectric permittivity at the vacuum–liquid interface on the boundary of emerging pores, the electrostrictive pressure behaves like saturated vapor inside the bubble and is directed outward from the nanopore surface. If this pressure is higher than the surface tension pressure, it displaces the liquid outward and the pore expands. In the opposite case, the nanopore collapses. The velocity of expansion of the nanopore is several times less than the speed of sound, but an order of magnitude greater than the velocity of fluid near the electrode caused by the ponderomotive forces. After the applied voltage is turned off, the negative pressure associated with electrical forces vanishes and the bubble collapses. In other words, the dynamics of the nanopore correspond to the dynamics of the Rayleigh bubble.

In chapter 6, we will generalize the theory of Zel'dovich–Fisher to the case of cavitation bubbles emerging in inhomogeneous pulsed electric fields.

## 1.6 A flat capacitor dipped in a dielectric fluid

A classic example of the action of ponderomotive forces is a capacitor partially dipped in dielectric fluid, as shown in figure 1.5. For simplicity, we assume that the distance between the capacitor plates $a$ is large enough that the surface tension of the liquid can be neglected: $a \gg 2\sigma_s/(\rho g)$, in which $g$ is the acceleration of free fall due to gravity [21].

The height of the liquid column can be estimated from the law of energy conservation. The capacitance of a capacitor partially filled with liquid is

$$C = \frac{\varepsilon_0 \varepsilon b}{a} h + \frac{\varepsilon_0 b}{a}(L - h) = \frac{\varepsilon_0 b}{a}(h \cdot (\varepsilon - 1) + L), \tag{1.24}$$

where $b$ is the width of the capacitor plates. The electrical energy stored in the capacitor is

$$W_C = \frac{1}{2} C U^2 = \frac{\varepsilon_0 b U^2}{2a}(h \cdot (\varepsilon - 1) + L), \tag{1.25}$$

here $U$ is the voltage on the electrodes.

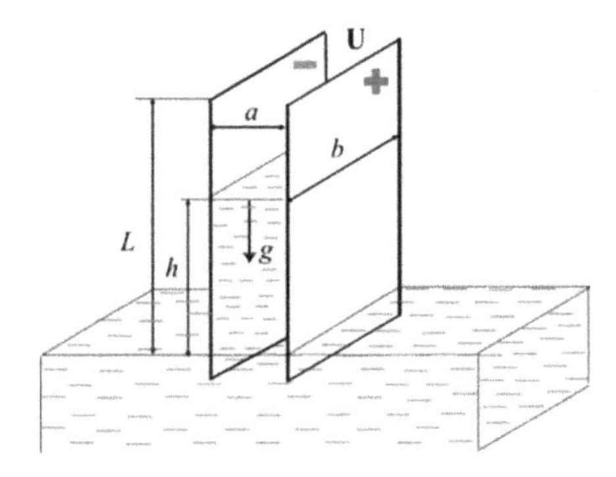


**Figure 1.5.** Flat capacitor dipped in a dielectric fluid.

The potential energy of the liquid column is

$$W_h = \frac{1}{2}\rho gabh^2. \tag{1.26}$$

The energy of the battery after charging the capacitor is

$$W_b = W_0 - CU^2. \tag{1.27}$$

Here $W_0$ is the initial energy stored in a battery and $CU^2$ is the energy which is spent on charging the capacitor. The total energy of the system consists of the energy of the capacitor, the potential energy of the liquid column and the energy of the battery:

$$\begin{aligned} W &= W_0 - CU^2 + \frac{1}{2}CU^2 + W_h \\ &= \frac{1}{2}\rho gabh^2 - \frac{\varepsilon_0 bU^2}{2a}(h \cdot (\varepsilon - 1) + L) + W_0. \end{aligned} \tag{1.28}$$

The height $h$ of the water column in the capacitor is determined by the minimum of the energy of the whole system ($\mathrm{d}W/\mathrm{d}h = 0$):

$$h = \frac{\varepsilon_0(\varepsilon - 1)U^2}{2\rho ga^2}. \tag{1.29}$$

For example, at $E = U/a = 2000$ kV m$^{-1}$, which is approximately 1.5 times smaller than the breakdown field in air under normal conditions, from (1.29) we obtain the height of the water column $h \approx 0.14$ m.

## 1.7 The polarization (Maxwell) time

The time for a dipole's reorientation in a polar dielectric fluid (water) $\tau_d$ is of the order of $10^{-12}$–$10^{-11}$ s [22]. It would seem that if the rise time of the voltage on the electrode is $\tau_0 \gg \tau_d$, the electric field (and induction) in the fluid would be determined by its dielectric permittivity. However, this is not true when the dielectric has a rather high conductivity. In this case, the currents of free charges in the dielectric lead to the neutralization of the space charge in it, and as a result cause a

**Table 1.2.** Conductances and polarization times for various types of water [24].

| Type of water | $\Sigma$ [S m$^{-1}$] | $\tau_M$ ($\varepsilon \approx 80$) [s] |
|---|---|---|
| Distilled | $5.5 \cdot 10^{-6}$ | $1.3 \cdot 10^{-4}$ |
| Potable (from a tap) | 0.005–0.05 | $1.4 \cdot 10^{-7} - 1.4 \cdot 10^{-8}$ |
| Sea | 2–5 | $1.4 \cdot 10^{-10}$ |

substantial decrease and displacement in the electric field. The neutralization time in a conductive liquid (electrolyte) can be estimated using the continuity equation for the volumetric charge density $\rho_f$, similar to how it is done in the theory of weakly ionized plasma [23]:

$$\frac{\partial \rho_f}{\partial t} + \nabla \cdot \vec{j}_f = 0, \tag{1.30}$$

where $\vec{j}_f = \Sigma \cdot \vec{E}$ is the current charge density in a fluid with conductivity $\Sigma$. Assuming $\Sigma = \text{const}$ and taking into account that $\nabla \cdot \vec{E} = \rho_f / \varepsilon\varepsilon_0$ from the Poisson equation (see chapter 4), (1.30) reduces to

$$\frac{\partial \rho_f}{\partial t} + \frac{\Sigma}{\varepsilon\varepsilon_0}\rho_f = 0. \tag{1.31}$$

The solution has the form:

$$\rho_f(t) = \rho_f(0) \cdot e^{-t/\tau_M}, \tag{1.32}$$

where $\tau_M = \frac{\varepsilon\varepsilon 0}{\Sigma}$ is the so-called polarization (Maxwell) time.

Table 1.2 lists the conductances and corresponding estimates of polarization times for various types of water.

If the voltage rise time on a needle-like electrode is $\tau_0 \sim 10^{-9} - 5 \cdot 10^{-9}$ s, then cavitation ruptures cannot form in seawater, because of the short $\tau_M$. This explains the inability of initiating sub-nanosecond discharge in salted water [25].

## 1.8 The flow induced in the vicinity of a needle-like electrode: a hydrostatic pressure

As we noted above, the development of discontinuities (cavitation) in water near a needle-like electrode requires the absolute value of the negative pressure to exceed a critical value of ~30 MPa. However, volumetric ponderomotive forces acting on the liquid cause it to flow toward the electrode, increasing the hydrostatic pressure, which compensates for the negative pressure.

We use linear approximation to estimate the displacement of the fluid and its velocity near the electrode. For simplicity, we assume that the electrode is a sphere of radius $r_{el}$. The equation of motion of an incompressible dielectric fluid under the action of volumetric Helmholtz force (see chapter 4) in a non-uniform electric field, neglecting viscosity, is

$$\rho\frac{\partial u}{\partial t} = -\frac{\partial p}{\partial r} - \frac{\varepsilon_0}{2}E^2\frac{\partial \varepsilon}{\partial r} + \frac{\varepsilon_0}{2}\frac{\partial}{\partial r}\left(E^2\rho\frac{\partial \varepsilon}{\partial \rho}\right), \tag{1.33}$$

where $p$ is the hydrostatic pressure and $\rho$ is the density of the liquid. For a polar liquid, $\frac{\partial \varepsilon}{\partial \rho}\rho = \alpha_E\varepsilon$ ($\alpha_E \approx 1.3-1.5$) [26], and for a non-polar liquid, $\rho\frac{\partial \varepsilon}{\partial \rho} = \frac{(\varepsilon-1)\cdot(\varepsilon+2)}{3}$ [26] (see chapter 4). Neglecting the inhomogeneity of $\varepsilon$ before the initiation of ruptures (the beginning of cavitation) in the liquid and considering the initial hydrostatic pressure $p \ll \frac{1}{2}\varepsilon_0 E^2\rho\frac{\partial \varepsilon}{\partial \rho}$, equation (1.33) reduces to

$$\rho_0\frac{\partial u}{\partial t} = \beta_E\frac{\partial E^2}{\partial r}. \tag{1.34}$$

Here, we introduced the variable $\beta_E = \frac{1}{2}\alpha_E\varepsilon_0\varepsilon$.

Assume, for example, a linear growth of the voltage on the electrode: $U = \frac{t}{\tau_0}U_0$. In this case, the field in the surrounding area is

$$E = \frac{t}{\tau_0}\cdot\frac{r_{el}}{r^2}U_0. \tag{1.35}$$

Substituting (1.35) into (1.34), we obtain the following equation for a local displacement of fluid $\zeta(r, t)$:

$$\frac{\partial^2\zeta}{\partial t^2} = -\frac{\beta_E}{\rho_0}\frac{t^2U_0^2}{\tau_0^2}\frac{r_{el}^2}{r^5} = -\frac{\beta_E}{\rho_0}\frac{t^2U_0^2}{r_{el}^2\tau_0^2}\frac{r_{el}^4}{r^5}. \tag{1.36}$$

Hence, the fluid velocity $u = \mathrm{d}\zeta/\mathrm{d}t$ at a time $t = \tau_0$ is equal to

$$u(r, \tau_0) = -\frac{\beta_E}{3\rho_0}\frac{U_0^2}{r_{el}^3}\left(1 - \frac{r_{el}^5}{r^5}\right)\frac{r_{el}^2}{r^2}\tau_0, \tag{1.37}$$

in which we took into account that the speed of fluid on the electrode is zero and the mass flux at $r > r_{el}$ is conserved: $4\pi r^2u = \mathrm{const}$. Then, from (1.37),

$$\zeta(r, \tau_0) = \frac{\beta_E}{12\rho_0}\frac{U_0^2}{r_{el}^3}\left(1 - \frac{r_{el}^4}{r^5}\right)\frac{r_{el}^2}{r^2}\tau_0^2 \approx -\frac{5\beta_E}{12\rho_0}\frac{U_0^2}{r_{el}^3}\frac{\delta r}{r_{el}}\tau_0^2, \tag{1.38}$$

where $\delta r = r - r_{ei} \ll r_{ei}$. Now, we will estimate the value $\tau_0$ at which the hydrostatic pressure associated with the compression of water becomes comparable to the electrostrictive pressure. Since the compression ratio of water is $K_W = 5\cdot 10^{-10}$ Pa$^{-1}$ [27], the ratio of the hydrostatic pressure associated with the compression of water to the electrostrictive pressure $P_E(\tau_0, r_{el}) = -\beta_E\frac{U_0^2}{r_{el}^2}$ is equal to

$$\frac{p}{|P_E|} \sim \frac{1}{K_w}\frac{|\zeta(\tau_0)|}{\delta r}\frac{1}{|P_E|} \approx \frac{5}{12\rho_0}\frac{1}{r_{el}^2K_w}\tau_0^2. \tag{1.39}$$

This implies that at the pulse front duration $\tau_0$ given by

$$\tau_0 \approx \tau_P = r_{el}\sqrt{\frac{12\rho_0 K_w}{5}} = \sqrt{\frac{12}{5}}\frac{r_{el}}{c_s} \approx 1.55\frac{r_{el}}{c_s}, \tag{1.40}$$

where $c_s = 1/\sqrt{\rho_0 K_w}$ is the speed of sound, the hydrostatic pressure near the electrode is $p \approx |P_E| \gg p_0$. So, at $\tau_0 \geqslant \tau_P$, the negative pressure is negligible and thus insufficient for the formation of discontinuities in the liquid. On the other hand, if $\tau_0 \ll \tau_P$, then $p \ll |P_E|$, therefore at $|p + P_E| \geqslant |P_{cr}|$ the development of cavitation becomes possible.

It follows from (1.37), that the maximum velocity $u_{max}$ is reached at a point $r/r_{el} = (7/2)^{1/5} \approx 1.3$ and it equals

$$|u(\tau_0)_{max}| \approx \frac{\beta_E}{3\rho_0}\frac{U_0^2}{r_{el}^3}\left(1 - \frac{2}{7}\right)\left(\frac{2}{7}\right)^{2/5}\tau_0 = 0.14\frac{|P_E(\tau_0, r_{el})|}{\rho_0 r_{el}}\tau_0, \quad \tau_0 \leqslant \tau_p. \tag{1.41}$$

In chapter 5, we will present the results of numeric calculations of the distributions of total pressure $p_{total} = p + P_E$ and velocity $\vec{u}$ in the vicinity of a needle-shaped electrode with a radius of curvature $r_{el} = 5$ μm, for a maximal voltage, $U_0 = 7$ kV, and at different rise times, $\tau_0$. For these parameters, the electrostriction pressure at the tip of the electrode is $P_E(\tau_0, r_{el}) = -220$ MPa. Substituting these values into (1.40) and (1.41), we find that the rise time of the voltage pulse $\tau_0 < 5$ ns is necessary for the occurrence of breaks in water and the corresponding absolute value of the maximal flow velocity is $|u(\tau_0)|_{max} \approx 33$ m s$^{-1}$. These simple estimates are in good agreement with the results we will present in chapter 5, which are based on more detailed calculations.

## References


[1] www.britannica.com/science/cavitation and www.amcsearch.com.au/facilities/ship-hydrodynamics/cavitation-research-laboratory/

[2] Lord Rayleigh O M F R S 1917 On the pressure developed in a liquid during the collapse of a spherical cavity *Phil. Mag.* **34** 94

[3] Zel'dovich Y B 1942 Theory of formation of a new phase. Cavitation *Zh. Eksp. Teor. Fiz* **12** 525

[4] Fisher J C 1948 The fracture of liquids *J. Appl. Phys.* **19** 1062

[5] Peebles P J E and Ratra B 2003 The cosmological constant and dark energy *Rev. Mod. Phys.* **75** 559

[6] Caupin F and Herbert E 2006 Cavitation in water: a review *C. R. Phys.* **7** 1000

[7] Kagan Yu 1960 The kinetics of boiling of pure liquid *J. Phys. Chem.* **34** 42

[8] Blander M and Katz J L 1975 Bubble nucleation in liquids *AIChE J.* **21** 833

[9] Shneider M N and Pekker M 2013 Cavitation in dielectric fluid in inhomogeneous pulsed electric field *J. Appl. Phys.* **114** 214906

[10] Vargaftic N B, Volkov B N and Voljak L D 1983 International tablets of the surface tension of water *J. Phys. Chem. Ref. Data* **12** 817

[11] Henderson S J and Speedy R J 1980 A Berthelot–Bourdon tube method for studying water under tension *J. Phys. E: Sci. Instrum.* **13** 778

[12] Hiro K, Ohde Y and Tanzawa Y 2003 Stagnations of increasing trends in negative pressure with repeated cavitation in water/metal Berthelot tubes as a result of mechanical sealing *J. Phys. D: Appl. Phys.* **36** 592
[13] Briggs L J 1950 Limiting negative pressure of water *J. Appl. Phys.* **21** 721
[14] Wurster C, Köhler M, Pecha R, Eisenmenger W, Suhr D, Irmer U, Brümmer F and Hülser D 1995 *Proc. 1st World Congress on Ultrasonics (Berlin)* ed Herbertz J (Duisburg: Universität Duisburg-Essen) part 1, p 635
[15] Greenspan M and Tschiegg C E 1967 Radiation-induced acoustic cavitation; apparatus and some results *J. Res. Natl Bur. Stand C* **71** 299
[16] Zheng Q, Durben D J, Wolf G H and Angell C A 1991 Liquids at large negative pressures: water at the homogeneous nucleation limit *Science* **254** 829
[17] Herbert E, Balibar S and Caupin F 2006 Cavitation pressure in water *Phys. Rev. E* **74** 041603
[18] Finch R D 1964 Influence of radiation on the cavitation threshold of degassed water *J. Acoust. Soc. Am.* **36** 2287
[19] Tolman R S 1949 The superficial density of matter at liquid–vapor boundary *J. Chem. Phys.* **17** 333
[20] Shneider M N and Pekker M 2016 Rayleigh scattering in the cavitation region emerging in liquids *Opt. Lett.* **41** 1090
[21] Landau L D and Lifshitz E M 1987 *Fluid Mechanics* (*Course of Theoretical Physics* vol 6) 2nd edn (Oxford: Butterworth-Heinemann)
[22] Both F 1951 The dielectric constant of water and the saturation effect *J. Chem. Phys.* **19** 391
[23] Raizer Yu P 1991 *Gas Discharge Physics* (Berlin: Springer)
[24] www.lenntech.com/applications/ultrapure/conductivity/water-conductivity.htm www.kayelaby.npl.co.uk/general_physics/2_7/2_7_9.html
Radiometer Analytical 2004 *Conductivity Theory and Practice* (Villeurbanne: Radiometer Analytical SAS)
[25] Starikovskiy A 2015 private communication
[26] Ushakov V Y, Klimkin V F and Korobeynikov S M 2005 *Breakdown in Liquids at Impulse Voltage* (Tomsk: NTL (in Russian))
Ushakov Y V, Klimkin V F and Korobeynikov S M 2007 *Impulse Breakdown of Liquids (Power Systems)* (Berlin: Springer)
[27] http://www.engineeringtoolbox.com/bulk-modulus-elasticity-d_585.html

**IOP** Publishing

Liquid Dielectrics in an Inhomogeneous Pulsed Electric Field

**M N Shneider and M Pekker**

# Chapter 2

## Classic cavitation

*The definition of cavitation and formulation of the basic problem. Cavitation in the subsonic flow of fluid in a pipe. Conditions for cavitation bubble formation near propeller blades. Cavitation generated by acoustic and shock waves. A new look at nucleation.*

### 2.1 The definition of cavitation and formulation of the basic problem

Cavitation is commonly defined as the formation of vapor cavities in a liquid (i.e. small liquid-free zones: 'bubbles' or 'voids') that are the consequence of forces acting upon the liquid. Usually, cavitation occurs when a liquid is subjected to rapid changes of pressure that cause the formation of cavities where the pressure is relatively low [1–3].

Those who have been in the mountains and tried boiling a kettle know that the boiling point of water decreases significantly with altitude. Figure 2.1 shows the dependence of the boiling point of ordinary drinking water on elevation above sea level [4]. The boiling temperature of distilled water is practically the same.

It would seem there is nothing surprising in the fact that by reducing pressure in the volume of the liquid one creates the conditions necessary for boiling. Yet, the fact is that if a liquid does not contain bubbles initially, they simply cannot form. Any bubble of a radius $R_b$ of the order of several nanometers will collapse under the forces of surface tension, because the forces per unit area associated with the curvature of the surface are inversely proportional to the radius of the bubble (Laplace pressure):

$$P_L = \frac{2\sigma_s}{R_b}. \tag{2.1}$$

The surface tension coefficient of water at a temperature of 20 °C for a planar boundary (or at a large radius of curvature) is $\sigma_s = 0.072$ N m$^{-1}$ [5]. Figure 2.2 shows the dependence of Laplace pressure on the radius of a bubble in water.

doi:10.1088/978-0-7503-1245-5ch2  

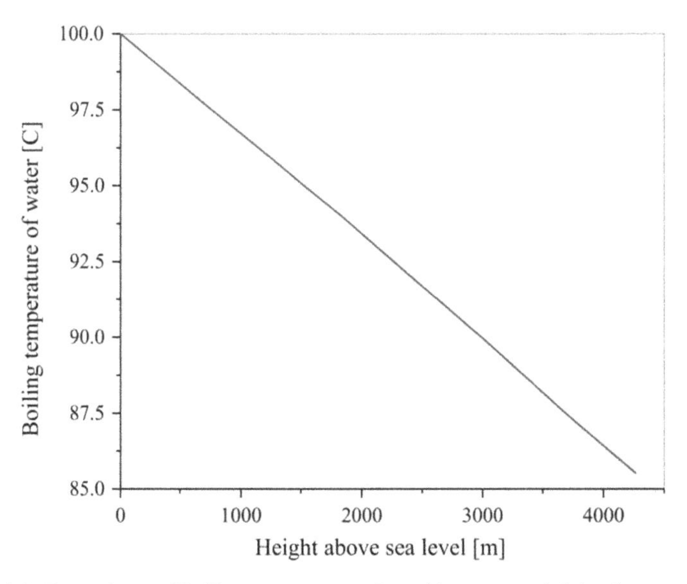


**Figure 2.1.** Dependence of boiling temperature of potable water on height above sea level.

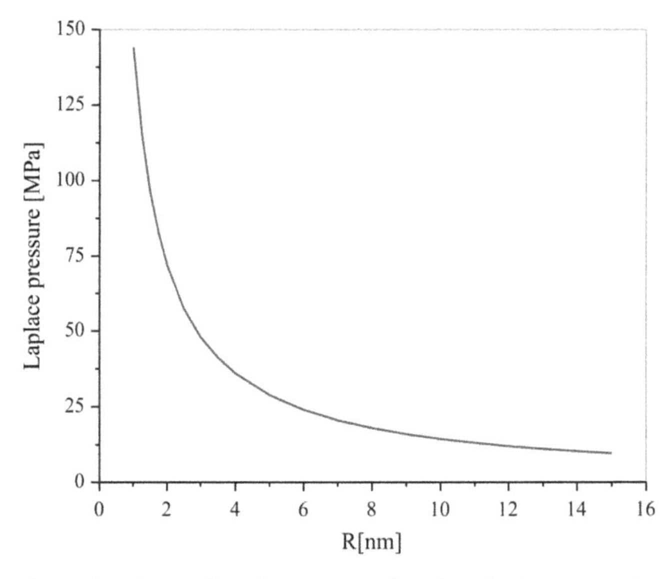


**Figure 2.2.** Dependence of surface tension of water per unit surface (Laplace pressure) on the radius of a cavity.

Since the vapor pressure in the bubble is roughly atmospheric ($P_{vap} \approx 0.1\,\text{MPa}$), equation (1) implies that the bubble's radius must be at least $R_b \approx 1.5\ \mu\text{m}$ to compensate the surface tension forces. At smaller radii, surface tension will collapse the bubble. Therefore, one can grow a bubble without additional heating by reducing the hydrostatic pressure in the liquid.

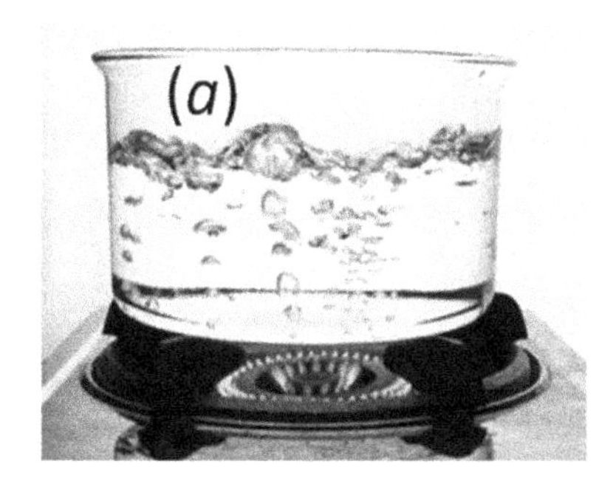

**Figure 2.3.** (a) Simmering stage: bubbles form on the surface of the vessel, then detach and rise while increasing in size. (b) Boiling stage: turbulent convection accompanies the formation of bubbles [8].

In accordance with (1.19) [6, 7], the number of fluctuations of the critical radius occurring per unit volume per unit time is

$$\frac{\mathrm{d}n_\mathrm{b}}{\mathrm{d}t} = \frac{3P^3}{16\pi\sigma_\mathrm{s}^3}\frac{k_\mathrm{B}T}{2\pi\hbar}\exp\left(-\frac{16\pi\sigma_\mathrm{s}^3}{3k_\mathrm{B}TP^2}\right). \tag{2.2}$$

At boiling temperature ($T = 373$ K) and a pressure of $P = P_\mathrm{vap} \approx 10^5$Pa, the exponent in (2.2) equals $16\pi\sigma_\mathrm{s}^3/3k_\mathrm{B}T \cdot P^2 = 1.8 \cdot 10^8$. This means that the probability of occurrence of a micron bubble due to thermal fluctuations is negligible.

But we know that if we put a kettle on a fire with highly pure distilled water, the water nonetheless boils at 100 °C and forms gas bubbles! The explanation for this 'amazing' fact is that the kettle surface always contains micropores that store gas, which facilitates the formation of microbubbles. The process of vigorous boiling is always preceded by a well-observed simmer stage, at which the bubbles that appear at the bottom rise to the surface and increase in size (figure 2.3(a)). The subsequent stage of intensive boiling, combined with the rapid growth in number of bubbles, is characterized by a developed turbulent convection (figure 2.3(b)) [8].

Microparticles and ions are always present in any, even highly purified and degassed, liquids. They serve as centers for the formation of micron-sized bubbles. However, in simmering and boiling the gas absorbed in the surface of the vessel plays the main role.

Later in this chapter, we will describe the processes of bubbles emerging due to thermal fluctuations, but first we will consider a few physical phenomena that result in cavitation.

## 2.2 Cavitation in the subsonic flow of fluid in a pipe

From the law of conservation of mass

$$\Theta = \rho S u = \mathrm{const} \tag{2.3}$$

follows

$$u = \frac{\Theta}{\rho S} = u_0\frac{S_0}{S}, \tag{2.4}$$

in which $\rho$ is the density of the fluid, $S = S(z)$ is the cross section area, $u = u(z)$ is the flow velocity at this cross-section, and $u_0$ and $S_0$ are the initial flow velocity and cross-section. From Bernoulli's law, the hydrostatic pressure in the cross section $S = S(z)$ is

$$P = P_0 + \rho\frac{u_0^2}{2} - \rho\frac{u^2}{2} = P_0 - \frac{u_0^2}{2}\left(\frac{S_0^2}{S^2} - 1\right). \tag{2.5}$$

Cavitation can form in the subsonic flow of a fluid in a tube of variable cross section (figure 2.4). When fluid enters the narrow part of the tube (section A–A) its speed increases and the pressure drops. At a sufficiently high ratio of cross sections between the pipe inlet and the neck, pressure $P$ may drop enough to cause the fluid to 'boil' between sections A–A and B–B. When the liquid flows from the narrow part to the expanding part of the tube (cross section B–B), the fluid velocity decreases, the hydrostatic pressure increases, and the bubbles collapse. The red line shows the cross section $S_1$, at which pressure becomes too high for bubbles to exist. Since the pressure of an ideal fluid in (2.5) cannot be negative, the 'boiling' in the tube always begins at the surface. If the neck cross section is too small, it follows from (5) that $P < 0$ and Bernoulli's equation does not apply.

## 2.3 Conditions for cavitation bubble formation near propeller blades

In the first chapter, we noted that propellers are a source of cavitation, which leads to intense destruction of the blades (figure 1.1(a)). The mechanism of cavitation formation in the vicinity of propeller blades differs from that described above for tubes with a variable cross section. A negative pressure emerges behind the propeller blades, which is associated with the centrifugal forces caused by the rotational motion of the fluid. We can estimate the magnitude of the negative pressure as

$$P_{-} \approx -\frac{\rho\omega_r^2 r^2}{2}, \tag{2.6}$$

where $\omega_r$ is the angular velocity of rotation of the rotor and $r$ is the radius of a point on the blade surface. Taking into account that at negative pressures $|P_{-}| > 30$ MPa the formation of cavitation is highly probable [9], we can estimate the angular velocity of

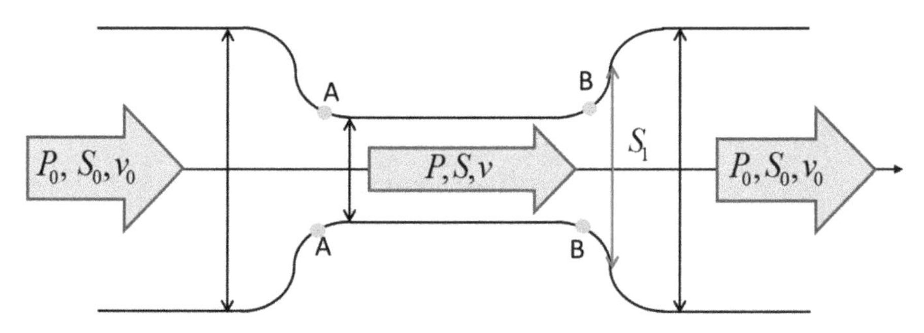


**Figure 2.4.** Generation of cavitation bubbles in a tube of variable cross section.

the propeller at which a region forms with a negative pressure sufficient to cause cavitation. Assuming, for example, that the radius of the blade $r \sim 1$ m, we obtain $\omega_r \approx \frac{1}{r}\sqrt{\frac{2|P_-|}{\rho}} \approx 250$ rad s$^{-1}$ from (2.6), or approximately 40 revolutions per second.

## 2.4 Cavitation generated by acoustic and shock waves

In a perfectly clean degassed liquid, cavitation in a traveling ultrasonic wave is practically impossible. Observations of the formation of such cavitation bubbles have been attributed solely to impurities (e.g. [10–12]). On the other hand, in the rarefaction phase of a standing ultrasonic wave of large amplitude, the tensile stress may become so large that cavitation occurs in the absence of impurities and dissolved gases.

Cavitation is also possible behind a shock wave, where the local negative pressure in a stretching region can exceed the required critical value (~−30 MPa in water). For example, one can generate such a shock wave by optical breakdown in a liquid.

Figure 2.5(a) shows a schematic diagram of an experiment of cavitation bubbles produced by a shock wave in water. In this experiment, an infrared nanosecond laser pulse with a wavelength of 1064 nm was focused in a microchannel that initiated optical breakdown in the water, creating a diverging shock wave [13]. When this wave reflected from the liquid–air interface, a region of negative pressure formed that was sufficient to form ruptures (nano- and microbubbles) in the water. A second nanosecond laser pulse of wavelength 532 nm with a controlled delay relative to the optical breakdown helped to observe the cavitation by scattering off the bubbles. Figure 2.5(b) shows a numerically calculated pressure distribution 58 ns after the initiation of the optical breakdown and 75 μm away from the water–air interface [13]. Blue shows the area of the negative pressure that occurred after the shock wave

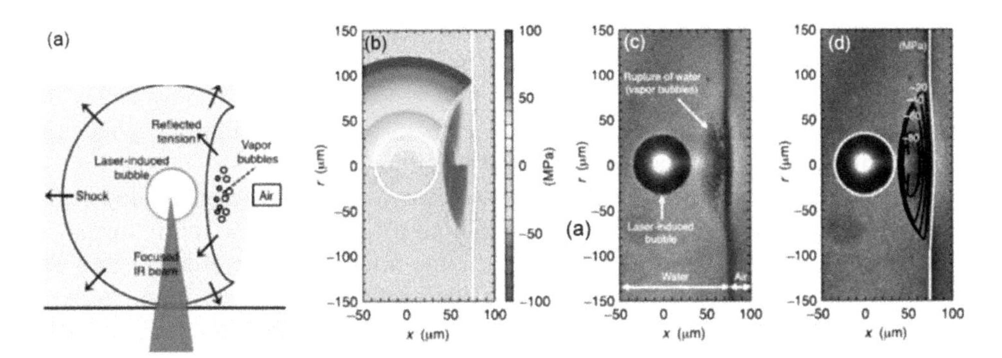


**Figure 2.5.** Rupture of air-saturated water. (a) Side view schematic of the experiment: interaction of a laser-induced shock with a free surface. (b) Simulated pressure distribution 58 ns after the optical breakdown and 75 μm away from the water–air interface surface. The upper and lower half planes show the distributions of the instantaneous pressure and the minimum pressure encountered during the computation, respectively. (c) An image of the nucleation of submicron cavitation vapor bubbles during the experiment. (d) Contours of minimum pressure and the simulated interface positions superimposed onto the image in (c). Reproduced with permission from [13]. Copyright 2012 the American Physical Society.

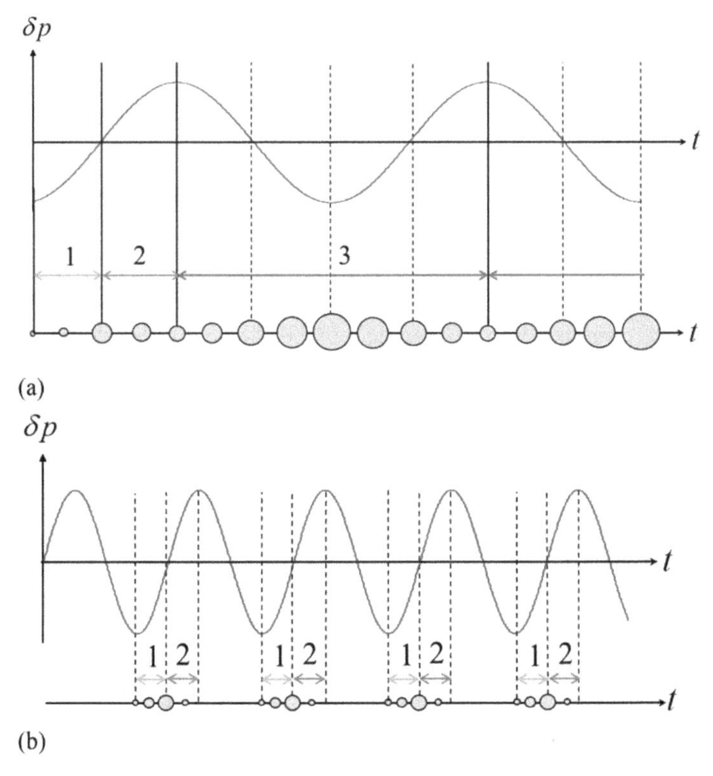


**Figure 2.6.** (a) Cavitation bubbles generated by a standing sound wave of low frequency: 1—bubble growth and its accumulation of more vapor; 2—bubble compression; and 3—bubble oscillation. (b) Generation of cavitation bubbles in a high frequency standing sound wave: 1—bubble growth and vapor accumulation, and 2—bubble compression and collapse.

reflected from the interface. Figure 2.5(c) shows scattering off the bubbles produced in the experiment by a 532 nm laser captured by a charge-coupled device camera with a 58 ns delay, just like in the simulation of figure 2.5(b). Figure 2.5(d) shows the superimposition of the calculated (b) and experimental (c) results, in which the contours mark the instantaneous pressure distribution.

When cavitation occurs in standing acoustic waves, the bubbles generated in the pressure 'trough' do not collapse at the maximum pressure if the sound frequency is sufficiently low. This is because, during the time interval of the order of half the period of the wave, bubbles have time to be filled with saturated vapor and grow to a large enough size such that their vapor pressure compensates the surface tension forces (Laplace pressure) and the excess pressure at the maximum wave (figure 2.6(a)). If the frequency of oscillation is too high, the bubbles that emerge at the pressure antinodes minima collapse during the succeeding pressure maxima (figure 2.6(b)), because the wave period is not long enough for them to grow to a size at which Laplace pressure no longer plays a significant role.

During the collapse of a bubble, the temperature in it can reach thousands or even tens of thousands of degrees Celsius, and the pressure thousands of times atmospheric pressure. This is due to the fact that, in the process of collapsing, the velocity of the bubble walls reach the speed of sound and consequently the pressure drop at the liquid–bubble interface in water reaches the value

$$\Delta P \sim \rho c_s^2 \approx 2.25 \cdot 10^3 \text{MPa}, \tag{2.7}$$

where $\rho$ and $c_s$ are the density and velocity of sound in the fluid [14–16]. The 'piston' that compresses the bubble is the Laplace pressure of (2.1), which is inversely proportional to the radius of the bubble.

At such high pressures and temperatures local shock waves form, excitation of atomic and molecular electronic levels takes place (the relaxation of which manifests as radiation), and chemical reactions initiate that are impossible under normal conditions. For these reasons the phenomenon of sonoluminescence occurs: light flashes appear at the collapse of cavitation bubbles, which are generated in an intensive standing ultrasonic wave in a liquid [17–19].

The conversion of spherical standing ultrasonic waves to light was already observed in the 1930s [20], but has long been completely incomprehensible. Only 60 years later did the appearance of fast cameras with high resolution allow physicists to associate sonoluminescence with the collapse of cavitation bubbles. Issues related to the dynamics of cavitation bubbles in acoustic fields, including sonoluminescence and the related sonochemistry (acoustic chemistry), have been well studied and are widely reflected in literature. Hence, we will not dwell on them and send the reader to the relevant reviews and monographs [21–26].

## 2.5 A new look at nucleation

The number of 'embryonic' voids of critical size arising per unit time per unit volume due to thermal fluctuations is given by formula (2.2). At water temperature $T = 300\,\text{K}$ and the coefficient of surface tension $\sigma_s = 0.072\ \text{Nm}^{-1}$, the critical pressure must be equal to $P_{cr} \approx 1200$ MPa. However, as we know from experiments (see table 1.1) the critical negative pressure is about 40 times smaller. Naturally, the question arises: what is the reason for such a strong discrepancy between theory and experiment? Equation (2.2) was obtained under the assumption of constant surface tension $\sigma_{0s}$, which corresponds to an infinitely thin boundary that separates the vacuum pore (or vapor bubble) from the liquid. However, as was first noted by Tolman [27], an approximation of the infinitely thin boundary is valid only when the thickness of the transition layer $\delta_b$ is much smaller than the pore radius $R_b$. Obviously, when the pore radius $R_b$ is of the order of $\delta_b$, the surface tension $\sigma_s$ should decrease and tend to zero at $R_b \to 0$. Tolman proposed the following empirical formula for the surface tension coefficient

$$\sigma_s = k_\sigma \sigma_{0s} \tag{2.8}$$

where

$$k_\sigma = 1/(1 + \delta_b/R_b). \tag{2.9}$$

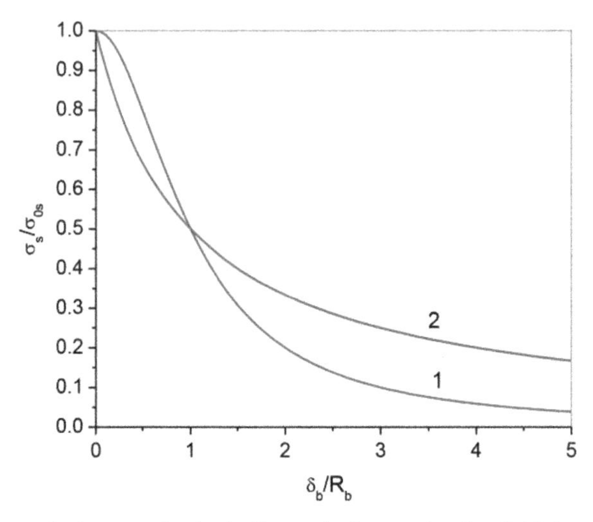


**Figure 2.7.** Dependence of $\sigma_s/\sigma_{0s}$ on $\delta_b/R_b$ in the 'Lorentzian' approximation (2.10) (curve 1) and the Tolman approximation (2.9) (curve 2).

Since at $\delta_b = 0$, the surface energy of the pore is $W_b = 4\pi\sigma_{0s}R_b^2$, the correction to that is related to the size of the transition layer $\delta_b$ and should be about $\delta W_b = -4\pi\sigma_{0s}\delta_b^2$. Based on this, we suggest expressing the approximation formula for the surface tension coefficient $\sigma_s$ in 'Lorentzian' form:

$$\sigma_s = \frac{\sigma_{0s}}{1 + \delta_b^2/R_b^2}. \tag{2.10}$$

Figure 2.7 shows the dependence $\sigma_s/\sigma_{0s}$ on $\delta_b/R_b$ for Tolman's (2.9) and the 'Lorentzian' (2.10) approximations.

The problem of nucleation in liquid helium at low temperatures and a given negative pressure was considered in [28]. In light of this work, the parameter $\delta_b$, introduced by Tolman phenomenologically, can be interpreted as the size of the layer at which there is a metastable equilibrium state between the gas and liquid. Since the surface tension for a gas is absent, when the radius of the fluctuations $R_b \to 0$ the corresponding surface tension $\sigma_s \to 0$. The dependence of the surface tension on the radius of curvature of the interphase transition region is still a matter of discussion (e.g. [29] and references therein).

Further, all the calculations will be carried out with the surface tension, taking into account the assumption (2.10). The energy required to create a bubble of radius $R_b$ in a liquid is equal to the work of the negative pressure $P_-$ plus the work against the forces of surface tension defined by the Laplace pressure $P_L = 2\sigma_s/r$ with the assumed adjustment for surface tension:

$$W(R_b) = -\int_0^{R_b} 4\pi r^2 |P_-| \mathrm{d}r + \int_0^{R_b} 8\pi r^2 \frac{\sigma_s}{r}\, \mathrm{d}r. \tag{2.11}$$

Substituting the value $\sigma_s$ from (2.10) we obtain

$$W(R_b) = -\frac{4\pi}{3}|P_-|R_b^3 + 4\pi R_b^2\sigma_{0s} - 4\pi\sigma_{0s}\delta_b^2 \ln\left(1 + \frac{R_b^2}{\delta_b^2}\right). \tag{2.12}$$

From (2.12) it is easy to find the value of the radius $R_{cr}$ at which the energy $W(R_b)$ reaches a maximum at a fixed value of $P_-$:

$$\begin{aligned} R_{cr} &= \frac{\sigma_{0s}}{|P_-|} + \sqrt{\left(\frac{\sigma_{0s}}{P_-}\right)^2 - \delta_b^2} = \frac{\sigma_{0s}}{|P_-|}\left(1 + \sqrt{1 - \frac{\delta_b^2 P_-^2}{\sigma_{0s}^2}}\right) \\ &= \frac{R_{cr,0}}{2}\left(1 + \sqrt{1 - \frac{4\delta_b^2}{R_{cr,0}^2}}\right), \quad R_{cr,0} = \frac{2\sigma_{0S}}{|P_-|} \end{aligned} \tag{2.13}$$

From (2.13) it follows that $R_{cr}$, depending on $\delta_b$, is within the range

$$\frac{1}{2} \leqslant \frac{R_{cr}}{R_{cr,0}} = \frac{1}{2}\left(1 + \sqrt{1 - \frac{4\delta_b^2}{R_{cr,0}^2}}\right) < 1. \tag{2.14}$$

The corresponding critical energy at this radius is

$$W_{cr} = -\frac{4\pi}{3}|P_-|R_{cr}^3 + 4\pi R_{cr}^2\sigma_{0s}\left(1 - \frac{\delta_b^2}{R_{cr}^2} \ln\left(1 + \frac{R_{cr}^2}{\delta_b^2}\right)\right). \tag{2.15}$$

Equation (2.15) can be rewritten as

$$\begin{aligned} \widetilde{W}_{cr} &= \frac{W_{cr}}{2\pi \cdot |P_-|R_{cr,0}^3/3} = \left(3\xi^2\left(1 - \frac{\delta_b^2}{R_{cr,0}^2\xi^2} \ln\left(1 + \frac{R_{cr,0}^2\xi^2}{\delta_b^2}\right)\right) - 2\xi^3\right), \\ \xi &= \frac{R_{cr}}{R_{cr,0}} = \frac{R_{cr}|P_-|}{2\sigma_0} \end{aligned} \tag{2.16}$$

The dependence of the parameters $\xi = R_{cr}/R_{cr,0}$ and $\widetilde{W}$ on $\delta_b/R_{cr}$ are shown in figure 2.8.

It is easy to see that at $|P_-| \ll \sigma_{0s}/\delta_b$,

$$R_{cr} \approx \frac{2\sigma_{0s}}{|P_-|} \quad \text{and} \quad W_{cr} \approx \frac{16\pi}{3}\frac{\sigma_{0s}^3}{P_-^2}, \tag{2.17}$$

which coincides with the values $R_{cr}$ and $W_{max}$ described by (1.11) and (1.12), respectively. The formula for the number of embryonic voids of critical radius produced per unit time per unit volume coincides with (2.2). The value $W_{cr} = 0$ at $\delta_b = 0.4785R_{cr,0}$ corresponds to the barrier-free cavitation that is not dependent on the fluid temperature (figure 2.8).

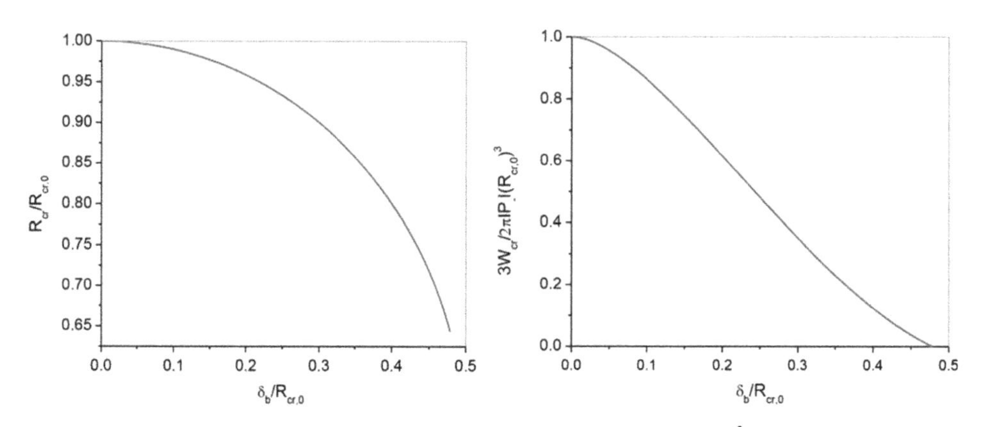


**Figure 2.8.** Dependence of the parameters $R_{cr}/R_{cr,0}$ and $3W_{cr}/(2\pi|P_{-}|R_{cr,0}^{3})$ on $\delta_b/R_{cr,0}$.

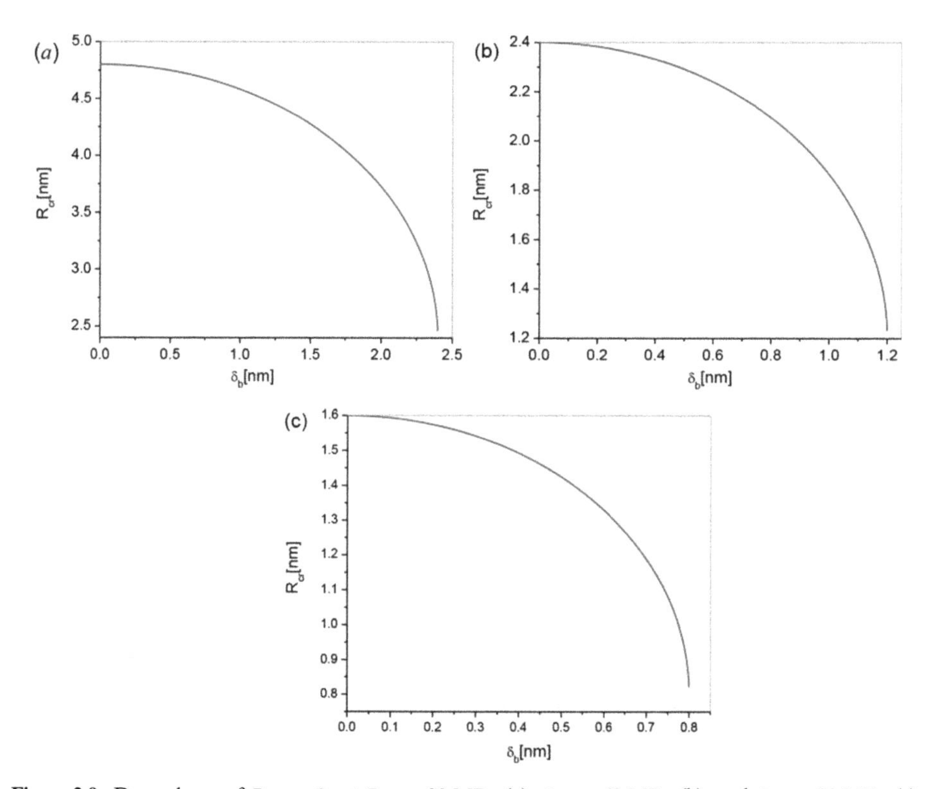


**Figure 2.9.** Dependence of $R_{cr}$ on $\delta_b$ at $P_{-} = -30$ MPa (a), $P_{-} = -60$ MPa (b), and $P_{-} = -90$ MPa (c).

The general expression for the nucleation rate is (see chapter 1)

$$\frac{\mathrm{d}n_b}{\mathrm{d}t} = \Gamma = \frac{3}{4\pi R_{cr}^3}\frac{k_B T}{2\pi\hbar}\exp\left(-\frac{W_{cr}}{k_B T}\right), \tag{2.18}$$

where $R_{cr}$ and $W_{cr}$ are given by expressions (2.13) and (2.15), respectively.

Figures 2.9 and 2.10 show the dependence of $R_{cr}$ and $\Gamma$ on $\delta_b$.

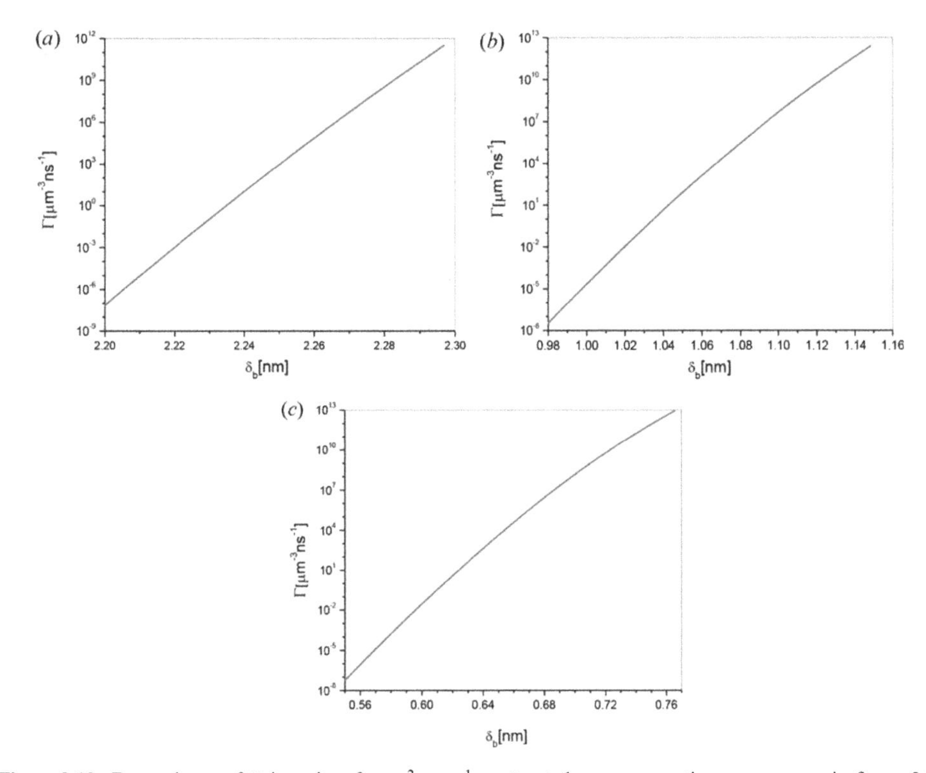


**Figure 2.10.** Dependence of $\Gamma$ in units of $\mu m^{-3} \cdot ns^{-1}$ on $\delta_b$ at the same negative pressures as in figure 2.9.

From figure 2.10 it follows that the number of nanopores produced within one $\mu m^3$ can reach $10^{11}$–$10^{13}$, that is an order of magnitude or more greater than the number of water molecules in this volume ($n_{water} \approx 3.3 \cdot 10^{10}\ \mu m^{-3}$). It is clear that this cannot be the case and, therefore, there must be a mechanism limiting the generation of nanopores.

Let us consider a possible mechanism that limits the generation of cavitation nanopores. We will assume that the volume in which a super-critical negative pressure occurs is of the order of $l_-^3$. The formation of nanopores is associated with an increased volume of the cavitation region, which in turn leads to excessive positive pressure that reduces the value of the negative pressure in the area.

The number of nanopores that appear in the negative pressure region $l_-^3$ over a characteristic time of pressure equilibration $\tau = l_-/c_s$ ($c_s$ is sound velocity) is equal to

$$N \approx l_-^3 \int_0^\tau \Gamma \, dt. \tag{2.19}$$

Accordingly, the relative change in volume of liquid in the area where nanopores are emerging is

$$\frac{\delta V}{V} \approx \frac{V_{cr} N}{l_-^3} = V_{cr} n_b, \tag{2.20}$$

where $V_{cr} = \frac{4}{3}\pi R_{cr}^3$ is the volume of a nanopore of the critical radius $R_{cr}$ ($R_{cr}$ corresponds to the initial negative pressure $P_{-,0}$) and $n_b$ is the density of these nanopores.

The value of the excess pressure $\delta P$ can be estimated from the simplest equation of state for a compressible fluid related to the sound velocity $c_s$:

$$\delta P = c_s^2 \delta\rho = c_s^2 \rho_0 \frac{\delta V}{V} = V_{cr} n_b c_s^2 \rho_0, \tag{2.21}$$

where $\rho_0$ is the unperturbed density of the fluid.

The absolute value of the total pressure in the bubble generation region is equal to

$$|P_-| = |P_{-,0}| - \delta P = |P_{-,0}| - V_{cr} n_b c_s^2 \rho_0. \tag{2.22}$$

Substituting $|P_-|$ into (2.15) we obtain

$$W_{cr} = W_{cr,0} + V_{cr}^2 c_s^2 \rho_0 n_b, \tag{2.23}$$

where

$$W_{cr,0} = -\frac{4\pi}{3}|P_-|R_{cr}^3 + 4\pi R_{cr}^2 \sigma_{0s}\left(1 - \frac{\delta_b^2}{R_{cr}^2}\ln\left(1 + \frac{R_{cr}^2}{\delta_b^2}\right)\right). \tag{2.24}$$

From (2.23), one can see that the increase in the number of pores per unit volume increases the value of $W_{cr}$ and thus reduces the rate of pore formation in (2.18).

Substituting (2.24) into (2.23) and then into (2.18) yields

$$\begin{aligned}\frac{dn_b}{dt} &= \frac{3}{4\pi R_{cr}^3}\frac{k_B T}{2\pi\hbar}\exp\left(-\frac{W_{cr,0}}{k_B T}\right)\cdot\exp\left(-\frac{c_s^2\rho_0 V_{cr}^2}{k_B T}n_b\right)\\ &= \Gamma(W_{cr,0})\exp\left(-\frac{c_s^2\rho_0 V_{cr}^2}{k_B T}n_b\right),\end{aligned} \tag{2.25}$$

where $\Gamma(W_{cr,0}) = \frac{3}{4\pi R_{cr}^3}\frac{k_B T}{2\pi\hbar}\exp(-\frac{W_{cr,0}}{k_B T})$.

If we recall that the connections of $R_{cr}$ and $W_{cr,0}$ with $\delta_b$ and $P_{-,0}$ are given by (2.13) and (2.15),

Equation (2.25) has a simple solution at fixed values $\delta_b$ and $P_-$:

$$n_{b,satur} = \frac{\ln\left(\Gamma(W_{cr,0})\frac{c_s^2\rho_0 V_{cr}^2}{k_B T}t + 1\right)}{\frac{c_s^2\rho_0 V_{cr}^2}{k_B T}}. \tag{2.26}$$

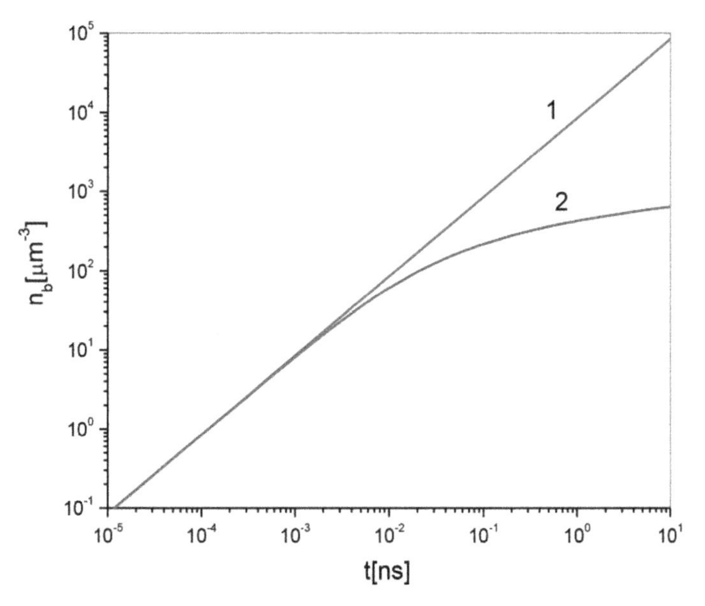


**Figure 2.11.** The dependence of the emerging nanopores' density on time. Line 1—without considering the effect of saturation (2.17); line 2—with saturation (2.25). $\delta_b = 2.255$ nm ($R_{cr} = 3.2$ nm), $|P_{-,0}| = 30$ MPa.

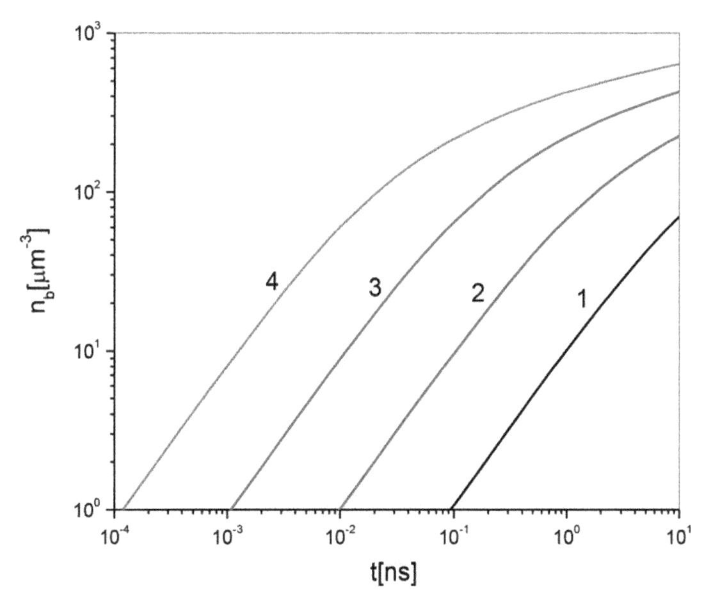


**Figure 2.12.** The dependence of the density of cavitation nanopores on time. Line 1—$\delta_b = 2.24$ nm ($R_{cr} = 3.262$ nm); line 2—$\delta_b = 2.245$ nm ($R_{cr} = 3.249$ nm); line 3—$\delta_b = 2.250$ nm ($R_{cr} = 3.235$ nm); line 4—$\delta_b = 2.255$ nm ($R_{cr} = 3.222$ nm). $|P_{-,0}| = 30$ MPa.

Figure 2.11 shows the dependence $n_b(t)$ with and without the saturation effect. For a small period of time when

$$\Gamma(W_{cr,0})\frac{c_s^2\rho_0 V_{cr}^2}{k_B T}t < 1, \tag{2.27}$$

the nanopore density increases linearly with time. Then, taking into account the saturation, the dependence $n_b(t)$ becomes logarithmic.

Figure 2.12 shows the dependence of $n_b(t)$ at various values of $\delta_b$, when the saturation effect is taken into account. The change of $\delta_b$ by 0.015 nm (0.7%) causes the change of the nanopores' density by an order, while the size of the nanopores remains almost the same, $R_{cr} \approx 3.2$ nm. Such a substantial threshold dependence of the density of the nanopores $n_b$ on $\delta_b$ allows one to determine the parameters $\delta_b$ and $R_{cr}$ for the known value of the critical negative pressure at which cavitation begins. An optical method for the detection of the nanopores' appearance at the initial stages of cavitation development will be considered in the following chapter.

The estimate (2.26) holds if $\Gamma(W_{cr,0})\tau = \Gamma(W_{cr,0})l_-/c_s \gg n_{b,satur}$.

It should be noted that the saturation effect of the nanopores' density occurs at any assumed dependence of the surface tension $\sigma_s(R_{cr})$.

## References


[1] Pearsall I S 1972 *Cavitation* (London: Mills and Boon)

[2] Brennen C E 1995 *Cavitation and Bubble Dynamics* (Oxford: Oxford University Press) http://authors.library.caltech.edu/25017/1/cavbubdynam.pdf

[3] Franc J-P and Michel J-M 2006 *Fundamentals of Cavitation* (*Fluid Mechanics and Its Applications* vol 76) (Berlin: Springer)

[4] American Society of Heating, Refrigerating and Air-Conditioning Engineers 1974 *ASHRAE Handbook—Fundamentals, Inch-Pound Edition* (Atlanta: ASHRAE)

[5] Haynes W M ed 2012 *CRC Handbook of Chemistry and Physics* 93th edn (Boca Raton, FL: CRC Press)

[6] Zel'dovich Y B 1942 Theory of formation of a new phase. Cavitation *Zh. Eksp. Teor. Fiz* **12** 525

[7] Fisher J C 1948 The fracture of liquids *J. Appl. Phys.* **19** 1062

[8] http://expertbeacon.com/act-quickly-and-calmly-when-suffering-boiling-water-burn/#.VwGVCqQrK7Q http://chefsblade.monster.com/benefits/articles/1250-7-crazy-kitchen-accidents?page=3

[9] Herbert E, Balibar S and Caupin F 2006 Cavitation pressure in water *Phys. Rev. E* **74** 041603

[10] Plesset M S 1969 The tensile strength of liquids *Cavitation State of Knowledge* (New York: ASME) 1–525

[11] Apfel R E 1970 The role of impurities in cavitation-threshold determination *J. Acoust. Soc. Am.* **48** 1179

[12] Gregorc B, Predin A, Fabijan D and Klasinc R 2012 Experimental analysis of the impact of particles on the cavitating flow *J. Mech. Eng.* **58** 238

[13] Ando K, Liu A-Q and Ohl C-D 2012 Homogeneous nucleation in water in microfluidic channels *Phys. Rev. Lett.* **109** 044501

[14] Gilmore F R 1952 The growth or collapse of a spherical bubble in a viscous compressible liquid *Technical Report* http://authors.library.caltech.edu/561/1/Gilmore_fr_26-4.pdf
[15] Plesset M and Prosperetti A 1977 Bubble dynamics and cavitation *Ann. Rev. Fluid Mech.* **9** 145
[16] Barber B P, Hiller R A, Löfstedt R, Putterman S J and Weninger K R 1997 Defining the unknowns of sonoluminescence *Phys. Rep.* **281** 65
[17] Crum C and Reynolds G T 1985 Sonoluminescence produced by stable cavitation *J. Acoust. Soc. Am.* **78** 137
[18] Barber B P and Putterman S 1991 Observation of synchronous picosecond sonoluminescence *Nature* **352** 318
[19] Gaitan D F, Crum C C, Churh C C and Roy R A 1992 Sonoluminescence and bubble dynamics for a single, stable, cavitation bubble *J. Acoust. Soc. Am.* **91** 3166
[20] Frenzel H and Schultes H Z 1934 Luminescence in ultra-ray layered water *Z. Phys. Chem.* B **27** 421
[21] Margulis M A 2000 Sonoluminescence *Phys. Usp.* **43** 259
[22] Ronald Young F 2000 *Sonoluminescence* (Boca Raton, FL: CRC Press)
[23] Hopkins S D, Putterman S J, Kappus B A, Suslick K S and Camara C G 2005 Dynamics of a sonoluminescing bubble in sulfuric acid *Phys. Rev. Lett.* **95** 254301
[24] Brenner M P, Hilgenfeldt S and Lohse D 2002 Single-bubble sonoluminescence *Rev. Mod. Phys.* **74** 425
[25] Margulis M A 2005 *Sonochemistry and cavitation* (Boca Raton, FL: CRC Press)
[26] Suslick K S and Flannigan D J 2008 Inside a collapsing bubble, sonoluminescence and conditions during cavitation *Ann. Rev. Phys. Chem.* **59** 659
[27] Tolman R S 1949 The effect of droplet size on surface tension *J. Chem. Phys.* **17** 333
[28] Lifshitz I M and Yu Kagan 1972 Quantum kinetics of phase transitions at temperatures close to absolute zero *Zh. Eksp. Teor. Fiz.* **62** 385
[29] Navascues G 1979 Liquid surfaces: theory of surface tension *Rep. Prog. Phys.* **42** 1131

**IOP** Publishing

Liquid Dielectrics in an Inhomogeneous Pulsed Electric Field

**M N Shneider and M Pekker**

# Chapter 3

# The physical properties of liquid dielectrics

*Dielectric liquids experience considerable strain under the influence of electrostrictive ponderomotive forces arising in pulsed inhomogeneous electric fields. Furthermore, the hydrostatic pressure reaches hundreds of atmospheres and greater, as in strong shock waves. Under these conditions, the compressibility of liquid dielectrics cannot be neglected, i.e. pressure and density must be related by the appropriate equation of state by taking into account compressibility and possible rarefaction. It is also obvious that the value of the static dielectric constant, as well as its value at optical frequencies, may differ greatly from the dielectric constant in the nanosecond range. Therefore, we find it necessary and useful to summarize published data on the equation of state, the dielectric constant, and the surface tension coefficients of liquid dielectrics. Of course, we do not intend and are not able to cover all the known data about physical properties of all liquid dielectrics; so we will limit our summary to the polar and non-polar liquids that pertain to examples considered in this book.*

## 3.1 Water

### 3.1.1 Equation of state

In contrast to ideal gas, the equation of state for liquids is impossible to derive theoretically from the statistical mechanics due to complex molecular interactions. Hence, the equation of state for liquids is introduced empirically.

Adiabatic processes (the absence of external heat supply and removal) occur in water at a practically constant temperature. This is explained by the specifics of the liquid molecular structure. Because of the high packing density, the fluid molecules experience repulsive forces in addition to the exchange of impulses during collisions caused by thermal motion. Pressure variation occurs substantially only due to changes in density from adiabatic compression or rarefaction. One of the generally accepted equations of state of water is the so-called barotropic Tait's approximation, proposed in 1888 [1]. At the core of Tait's approach lies a simple differential equation relating the pressure $p$ to the density $\rho$ of water:

$$\frac{n_{\mathrm{w}}}{\rho}\left(\frac{\partial \rho}{\partial p}\right)_T = \frac{1}{B + p}, \tag{3.1}$$

where $B$ and $n_{\mathrm{w}}$ are parameters determined from measurements. Integrating equation (3.1), we obtain the Tait equation of state

$$\left(\frac{\rho}{\rho_0}\right)^{n_{\mathrm{w}}} = \left(\frac{B + p}{B + p_0}\right). \tag{3.2}$$

For fresh water, $\rho_0 = 1000$ kg m$^{-3}$ is the density of water at a pressure of $p_0 = 1$ bar ($10^5$ Pa), $n_{\mathrm{w}} = 7.15$ and $B = 3.072 \cdot 10^8$ Pa. The dependences of parameters $n_{\mathrm{w}}$, $B$ and $\rho_0$ on the temperature and salinity of water can be found in [2, 3].

From the definition of the square of the speed of sound in water,

$$c_{\mathrm{s}}^2 = \left(\frac{\partial p}{\partial \rho}\right)_T \approx \text{const} \tag{3.3}$$

follows a linear (or acoustic) approximation for the equation of state:

$$p - p_0 = c_{\mathrm{s}}^2(\rho - \rho_0). \tag{3.4}$$

Figure 3.1 compares this linear approximation at $c_{\mathrm{s}} = 1482$ m s$^{-1}$ to Tait's equation (3.2).

Since the critical negative pressure for cavitation associated with electrostrictive forces is about –30 MPa, it is natural to expect that the hydrostatic pressure, which rises to compensate the electrostrictive forces, does not exceed 50–100 MPa (see

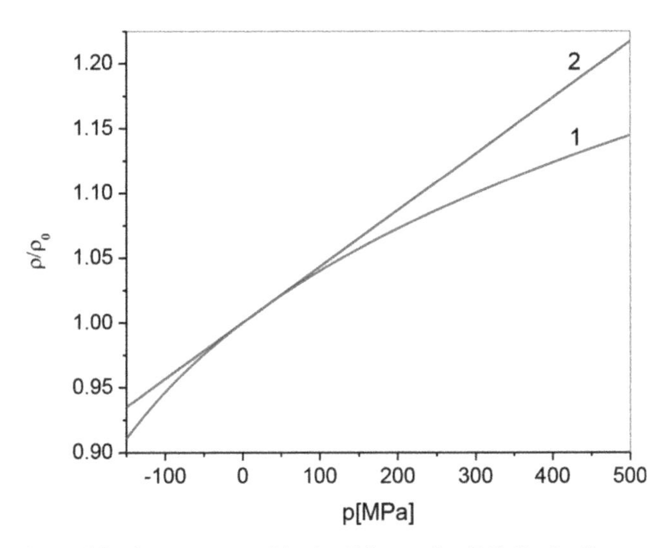


**Figure 3.1.** Dependence of density on pressure. Line 1—Tait equation (3.2); line 2—linear approximation (3.4).

chapters 4 and 5). Therefore, the linear acoustic approximation for the equation of state (3.4) can be used with an accuracy of about 1% at both negative and positive pressures in the [–100,100] MPa range.

It should be noted that other more detailed and sophisticated equations of state have been proposed for water (e.g. [4–6]). However, if heat release is insignificant and the temperature remains practically unchanged, as happens in the pre-breakdown stages of impulse breakdown in water, one can use either the Tait equation of state (3.2) or the linear acoustic approximation (3.4) for numerical calculations.

### 3.1.2 The dielectric constant of water

The dielectric constant of water depends on the electromagnetic wave frequency and is a complex function $\varepsilon = \varepsilon' + i\varepsilon''$. The following expressions determine the corresponding values of the refractive index $n$ and the attenuation coefficient $\kappa$ [7]:

$$n = \left(\left(\sqrt{\varepsilon'^2 + \varepsilon''^2} + \varepsilon'\right)/2\right)^{1/2}, \quad \kappa = \left(\left(\sqrt{\varepsilon'^2 + \varepsilon''^2} - \varepsilon'\right)/2\right)^{1/2}. \tag{3.5}$$

The dielectric constant of water can be approximated as

$$\varepsilon(\nu) = \frac{\varepsilon_1 - \varepsilon_2}{1 + i\omega\tau_1} + \frac{\varepsilon_2 - \varepsilon_3}{1 + i\omega\tau_2} + \varepsilon_3, \qquad \omega = 2\pi\nu \tag{3.6}$$

up to frequencies $\nu$ of several hundred gigahertz [8]. Table 3.1 shows the values of $\varepsilon_1$, $\varepsilon_2$, $\varepsilon_3$, $\tau_1$ and $\tau_2$ at different temperatures [8].

Figure 3.2 shows the dependence of the real and imaginary parts of the dielectric constant of water $\varepsilon$ on frequency at a temperature of 25 °C and corresponding values of the index of refraction $n$ and the attenuation coefficient $\kappa$, computed using (3.5). At $\nu < 5$ GHz, the dielectric constant has a weak frequency dependence. Therefore, when the voltage rise time on the electrode is longer than ~0.2 ns, the real part of the dielectric constant can be considered as a constant.

The dielectric constant of water depends not only on the frequency of the electromagnetic waves, but also on the value of the electric field. It is very difficult to

**Table 3.1.** Dielectric relaxation parameters of water.

| $T$ [°C] | $\varepsilon_1$ | $\tau_1$ [ps] | $\varepsilon_2$ | $\tau_2$ [ps] | $\varepsilon_3$ |
|---|---|---|---|---|---|
| 0.2 | 87.57 | 17.67 | 6.69 | 0.9 | 3.92 |
| 5 | 85.89 | 14.92 | 6.76 | 1.0 | 4.10 |
| 10 | 83.93 | 12.70 | 6.57 | 0.9 | 4.08 |
| 15 | 82.24 | 11.00 | 6.64 | 1.0 | 4.34 |
| 20 | 80.31 | 9.60 | 6.53 | 1.2 | 4.42 |
| 25 | 78.32 | 8.38 | 6.32 | 1.1 | 4.57 |
| 30 | 76.39 | 7.39 | 5.75 | 0.9 | 4.60 |
| 35 | 74.91 | 6.69 | 6.22 | 1.5 | 4.74 |

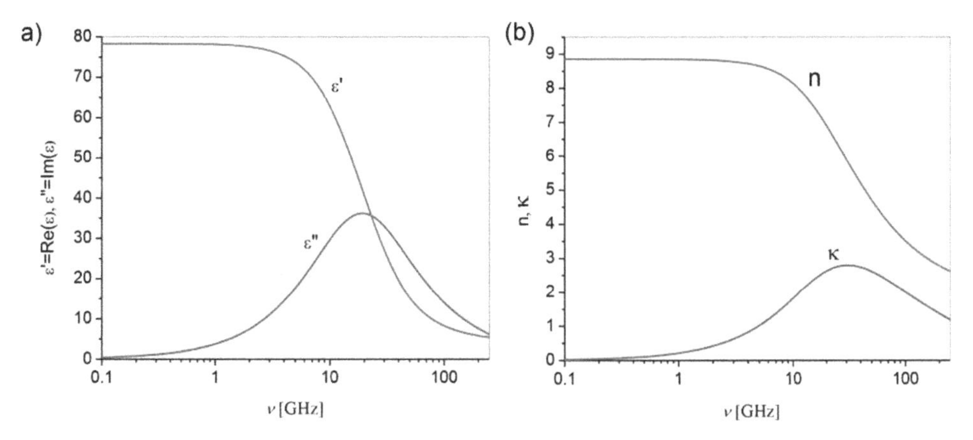


**Figure 3.2.** Dependences of water parameters on microwave frequency: real and imaginary dielectric constant values of water (a), and its index of refraction $n$ and the attenuation coefficient $\kappa$ (b) at a temperature of $T = 25$ °C.

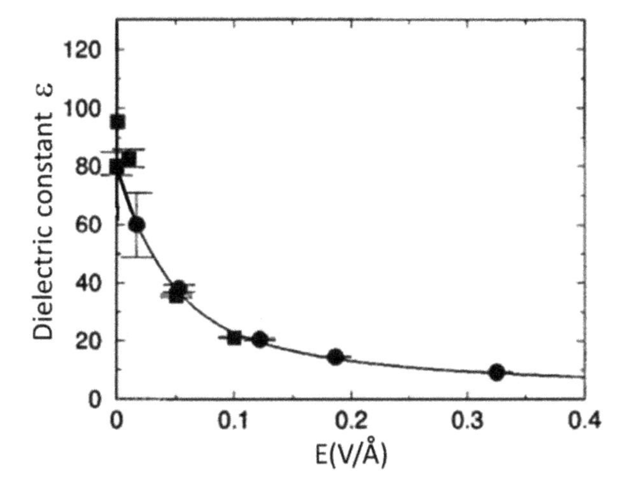


**Figure 3.3.** Comparison between results from different simulations and a theoretical prediction for the dependence of the dielectric constant of water on a strong electric field. The solid curve shows the theoretical prediction from [11], and the circles and squares show the molecular dynamics simulation results obtained by different approaches in [11] and [12], respectively. Reproduced with permission from [11]. Copyright 1999 AIP Publishing LLC.

experimentally study the dielectric constant in large electric fields when the alignment of polar molecules becomes significant. Even at field strengths of $10^7$–$10^8$ V m$^{-1}$, a breakdown develops in the water in a time of the order of milliseconds [9]. Therefore, the theoretical and computational data, in particular those produced by molecular dynamics methods, are of great importance. Figure 3.3 shows the values of the dielectric constant of water at high fields derived theoretically on the basis of the Onsager–Kirkwood theory [10] and from some numerical calculations by molecular dynamics methods [11, 12].

**Table 3.2.** Dependence of the static dielectric constant of water and steam on pressure and temperature.

| $p$ [MPa] | 0 °C | 25 °C | 50 °C | 75 °C | 100 °C | 150 °C |
|---|---|---|---|---|---|---|
| 0.1 | 87.81 | 78.46 | 69.91 | 62.24 | 1 | 1 |
| 1 | 87.83 | 78.47 | 69.92 | 62.25 | 55.43 | 43.95 |
| 5 | 88.05 | 78.65 | 70.09 | 62.42 | 55.59 | 44.12 |
| 10 | 88.28 | 78.85 | 70.27 | 62.59 | 55.76 | 44.30 |
| 50 | 90.07 | 80.36 | 71.66 | 63.93 | 57.08 | 45.67 |
| 100 | 92.04 | 82.08 | 73.22 | 65.42 | 58.55 | 47.14 |
| 150 | 93.71 | 83.57 | 74.62 | 66.74 | 59.82 | 48.40 |
| 200 | 95.20 | 84.94 | 75.89 | 67.95 | 61.00 | 49.54 |
| 300 | 97.69 | 87.34 | 78.17 | 70.14 | 63.10 | 51.55 |
| 500 | 101.42 | 91.16 | 81.84 | 73.69 | 66.57 | 54.85 |

The dielectric constant of water, as well as of any other polar liquid, is to a large extent determined by the reorientation of the molecules in an electric field and the dipole–dipole interactions of nearest neighbors. These processes strongly depend on the temperature of the fluid and its density (pressure) (see chapter 4). Table 3.2 presents data of the dependence of the dielectric constant of water on temperature and pressure [13].

### 3.1.3 The surface tension of water

Figure 3.4 shows the dependence of the surface tension, the saturated vapor pressure and the dynamic viscosity of water on temperature at atmospheric pressure [14].

The surface tension values in figure 3.4 are obtained for a planar, or large radius of curvature, water–gas boundary, which is substantially greater than the thickness of a transition layer at the gas–liquid interface. To determine the surface tension of water droplets or bubbles, with size of the order of or less than 10–20 nm, it is necessary to consider a correction to the planar surface tension coefficient that depends on the radius of curvature of the interface boundary (for details see chapter 2).

## 3.2 Experimental data related to oil and some other liquid dielectrics

### 3.2.1 Speed of sound and equation of state

The speed of sound in oil and other liquid dielectrics under normal conditions is not much different from the speed of sound in water. Table 3.3 shows sound speed values for some polar and non-polar liquids at temperatures of 20–25 °C [15].

Earlier, we considered the approximate Tait's adiabatic equation of state of water (3.2) and the linear acoustic approximation (3.4). In this book, we will not go into the details of the equation of state of oils and other liquids. Since we are interested in a range of subcritical and critical conditions for the development of cavitation, which occurs at rather small negative pressures $|P_{-}| \ll 100$ MPa [16, 17], we expect that the change of hydrostatic pressure is of a similar scale. Therefore, the acoustic

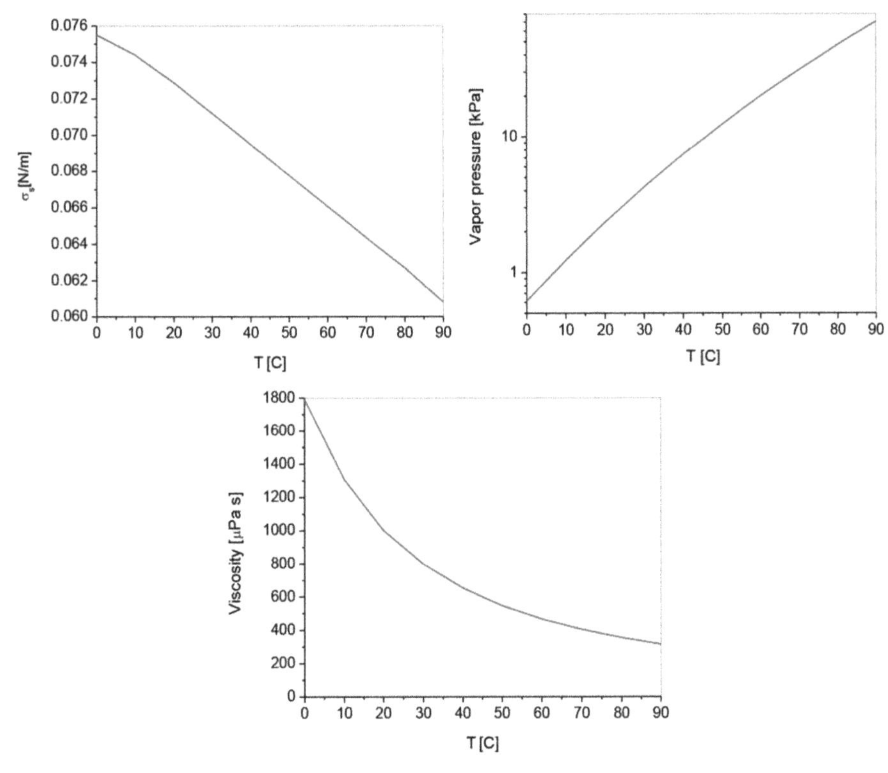


**Figure 3.4.** Dependence of surface tension, saturated vapor pressure and dynamic viscosity of water on temperature at atmospheric pressure.

**Table 3.3.** The velocity of sound in some common liquids.

| Liquid | $c_s$ [m s$^{-1}$] ($T$ = 20–25 °C) |
|---|---|
| Water | 1482 |
| Sea water | 1522 |
| Castor oil | 1490 |
| Lubrication oil | 1461 |
| Acetone ($CH_3$—C(O)—$CH_3$) | 1170 |
| Glycerol $C_3H_5(OH)_3$ | 1920 |
| Alcohol, ethyl ($C_2H_5OH$) | 1144 |

approximation (3.4) with the appropriate sound velocity can be used as the equation of state for oils and other dielectric fluids for which the speed of sound is close to the speed of sound in water. But this assumption requires experimental verification.

**Table 3.4.** Dielectric constant $\varepsilon$ and dynamic viscosity $\eta$ at $10^{-3}$ Pa · s for various liquids.

| Liquid | $\varepsilon$ ($T$ = 20 °C) | $\eta$ in$10^{-3}$[Pa · s] |
|---|---|---|
| Transformer oil | 2.1–2.4 | 15–24 ($T$ = 20 °C) |
| Capacitor oil | 2.1–2.3 | 27–41 ($T$ = 20 °C) |
| Castor oil | 4.0–4.5 | 125 ($T$ = 50 °C) |
| Cable oil | 2.1–2.4 | 720 ($T$ = 20 °C) |
| Alcohol, ethyl ($C_2H_5OH$) | 25.3 | 1.074 ($T$ = 25 °C) |
| Glycerol $C_3H_5(OH)$ | 46.51 | 934 ($T$ = 25 °C) |
| Acetone $CH_3$–C(O)–$CH_3$ | 21 | 0.306 ($T$ = 25 °C) |

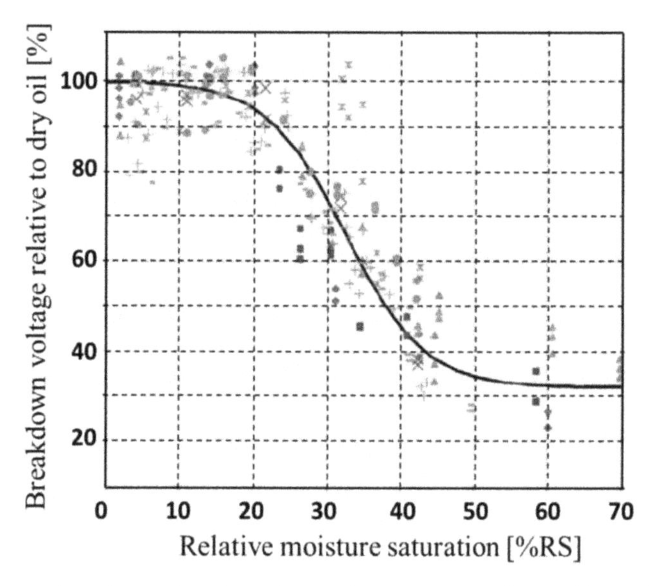


**Figure 3.5.** Measured dependency of the breakdown voltage on the relative moisture saturation for eight different types of mineral and synthetics oils. The solid curve corresponds to averaging over all measurements. Reproduced courtesy of Vaisala [18].

### 3.2.2 Dielectric constant

Table 3.4 lists the dielectric constants of some oils [18] and undiluted ethyl alcohol, glycerol, and acetone [14].

The breakdown voltage in oil strongly depends on the water content in it. Figure 3.5 shows measurements of the dependence of the relative breakdown voltage on the humidity for different kinds of oil [18].

### 3.2.3 Surface tension

Table 3.5 shows values of surface tension coefficients of some oils at the interface between air and water at atmospheric pressure and the temperature 20 °C [19].

**Table 3.5.** Surface tension of oils at the interface between air and water at $p$ = 1 atm and $T$ = 20 °C.

| Oil | $\sigma_{0s}$ [N m$^{-1}$] (oil–air) | $\sigma_{0s}$ [N m$^{-1}$] (oil–water) |
|---|---|---|
| Perfume | 0.0302 | 0.0438 |
| Transformer | 0.0306 | 0.0424 |
| Engine | 0.0356 | 0.0338 |
| Cable | 0.0367 | 0.0271 |

## 3.3 Liquid helium

From the wide variety of experimental data relating to cryogenic liquids, we present only some of the data relating to liquid helium, which we consider in this book.

### 3.3.1 Equation of state

The equation of state of liquid helium-4 at temperatures close to 0 K is described by the empirical formula [20]:

$$p = A_0(\rho - \rho_0) + A_1(\rho - \rho_0)^2 + A_2(\rho - \rho_0)^3, \tag{3.7}$$

where $A_0$ = 56 [Mpa kg m$^{-3}$], $A_1$ = 1 097 [Mpa kg$^2$ m$^{-6}$] and $A_2$ = 7 330 [Mpa kg$^3$ m$^{-9}$].

The expression for speed of sound follows from (3.7):

$$c_s = \left(\left(\frac{\partial p}{\partial \rho}\right)_T\right)^{1/2} = A_0 + 2A_1(\rho - \rho_0) + 3A_2(\rho - \rho_0)^2. \tag{3.8}$$

One can determine the dependence of the pressure $p$ of liquid helium on density $\rho$ from the known dependence of sound velocity on pressure at a given temperature. For example, [21] provides experimental dependences of the sound velocity on atmospheric pressure at $T = 0.115$ K in helium-3 and helium-4. The corresponding approximation formulas for the pressure derivatives $(\partial p/\partial \rho)_T$ are

$$\left(\frac{\partial p}{\partial \rho}\right)_T = c_{\mathrm{He}^4}^2 = \left(14.3 \cdot \left(p + 9.5 \cdot 10^5\right)\right)^{2/3} \tag{3.9}$$

for helium-4 and

$$\left(\frac{\partial p}{\partial \rho}\right)_T = c_{\mathrm{He}^3}^2 = \left(19.23 \cdot \left(p + 3 \cdot 10^5\right)\right)^{2/3} \tag{3.10}$$

for helium-3. Figure 3.6 shows these, as well as (8) at $T \to 0$, for helium-4.

Integrating (3.7) and (3.8), and taking into account that the densities of helium-4 and helium-3 at pressures close to atmospheric are $\rho_{\mathrm{He}^4} \approx 145$ kg m$^{-3}$ [14]

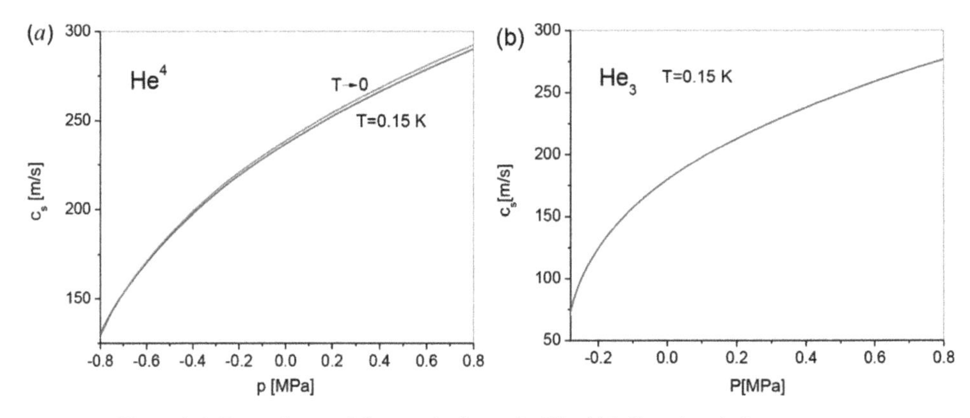


**Figure 3.6.** Dependence of the speed of sound of liquid helium-4 and -3 on pressure.

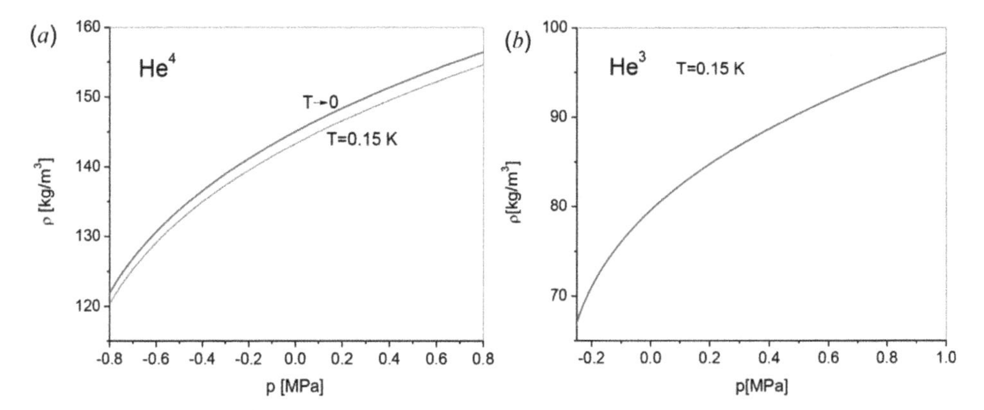


**Figure 3.7.** Dependence of density on pressure for liquid helium-4 and helium-3. The curves $T = 0.15$ K and $T \to 0$ in plot (a) correspond to (11) and (7), respectively.

and $\rho_{\mathrm{He}^3} \approx 82\ \mathrm{kg\ m^{-3}}$ [22], we obtain the corresponding approximate equations of state:

$$p = 7.57 \cdot (\rho - 93.24)^3 - 9.5 \cdot 10^5, \ (\mathrm{He}^4) \tag{3.11}$$

$$p = 13.7 \cdot (\rho - 51.55)^3 - 3 \cdot 10^5, \ (\mathrm{He}^3) \tag{3.12}$$

in which pressure $p$ is in pascals. Figure 3.7 shows the corresponding dependence of density on pressure.

Figure 3.8 shows the dependence of sound velocity in helium-4 on temperature at atmospheric pressure. The kink in the curve at $T = T_\lambda = 2.172$ K corresponds to the phase transition of helium-4 from the normal to the superfluid state [23].

Figure 3.9 shows similar data at different pressures [20].

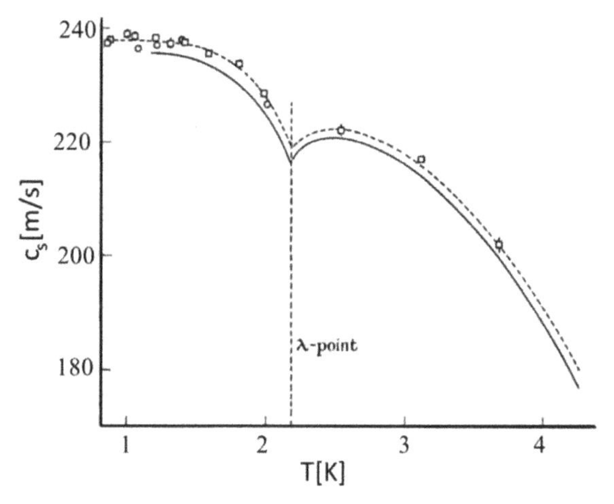


**Figure 3.8.** Dependence of sound velocity in helium-4 on temperature at atmospheric pressure. The solid line corresponds to earlier results. The kink in the curve corresponds to the phase transition of helium-4 from the normal ($T > T_\lambda$) to the superfluid state ($T < T_\lambda$) [23].

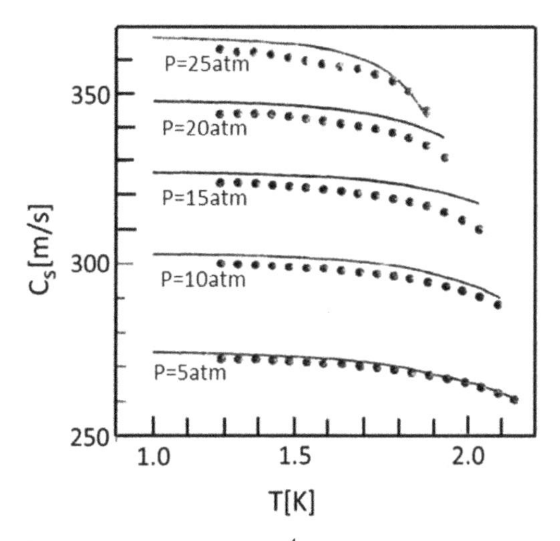


**Figure 3.9.** Dependence of sound velocity in liquid $He^4$ on temperature at different pressures. The dots and curves represent experimental [24] and computed [20] results, respectively. Reproduced from [20]. Copyright 1977 AIP Publishing LLC.

### 3.3.2 Dielectric constant

Table 3.6 lists the dependence of the dielectric constant on temperature for liquid helium-4 [14].

### 3.3.3 Surface tension

Finally, figure 3.10 shows the dependence of the surface tension of liquid helium on temperature [14, 25].

**Table 3.6.** Dielectric constant and density of liquid $He^4$ at different temperatures [14].

| $T$ (K) | $\varepsilon$ | $\rho$[kg m$^{-3}$]($P$ = 0.1 Mpa) |
|---|---|---|
| 0.0 | 1.057225 | 145.1397 |
| 0.5 | 1.057254 | 145.1377 |
| 1.0 | 1.057246 | 145.1183 |
| 1.5 | 1.057265 | 145.1646 |
| 2.0 | 1.057449 | 145.6217 |
| 2.5 | 1.057135 | 144.842 |
| 3.0 | 1.055683 | 141.2269 |
| 3.5 | 1.053615 | 136.0736 |
| 4.0 | 1.050770 | 128.9745 |
| 4.5 | 1.046725 | 118.8553 |

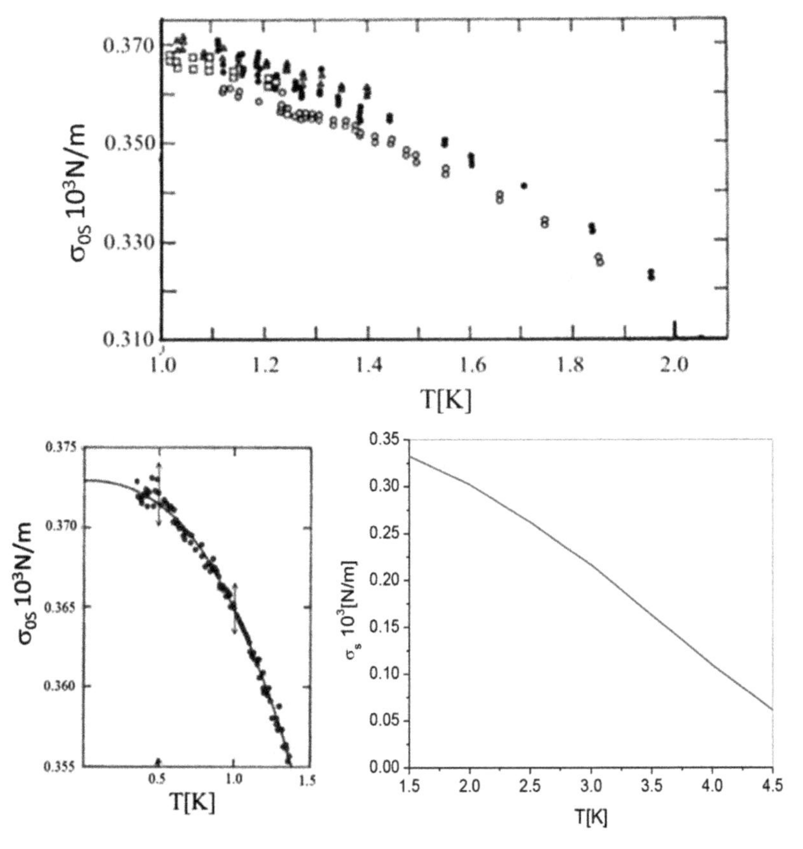


**Figure 3.10.** Surface tension of liquid helium in different temperature ranges. Top and bottom left panels reproduced from [25]. Copyright 1965 the American Physical Society.

## References


[1] Tait P G 1965 Report on some of the physical properties of fresh water and sea water *Report on the Scientific Results of the Voyage of the HMS Challenger During the Years 1873–76* (New York: Johnson Reprint Corp)

[2] Li Y H 1967 Equation of state of water and sea water *J. Geophys. Res.* **72** 2665

[3] Nigmatulin R I and Bolotnova R 2003 The equation of state of liquid water under static and shock compression *Proc. 4th Zababakhin Scientific Readings* (Snezhinsk: Russian Federal Nuclear Center)

[4] Nigmatulin R I and Bolotnova R 2008 Wide-range equation of state for water and steam: calculation results *High Temp.* **46** 325

[5] Hayward A T J 1967 Compressibility equations for liquids: a comparative study *Brit J. Appl. Phys.* **18** 965

[6] Gurtman G A, Kirsch J W and Hastings C R 1971 Analytical equation of state for water compressed to 300 Kbar *J. Appl. Phys.* **42** 851

[7] Landau L D and Lifshitz E M 1991 *Electrodynamics of Continuous Media (A Course of Theoretical Physics* vol 8*)* 3rd edn (Oxford: Pergamon)

[8] Buchner R, Barthel J and Stauber J 1999 The dielectric relaxation of water between 0°C and 35°C *Chem. Phys. Lett.* **306** 57

[9] Ushakov V Y, Klimkin V F and Korobeynikov S M 2007 *Impulse Breakdown of Liquids (Power Systems)* (Berlin: Springer)

[10] Both F 1951 The dielectric constant of water and the saturation effect *J. Chem. Phys.* **19** 391

[11] Yeh I-C and Berkowitz M L 1999 Dielectric constant of water at high electric fields: molecular dynamics study *J. Chem. Phys.* **110** 7935

[12] Sutmann G 1998 Structure formation and dynamics of water in strong external electric fields *J. Electroanal. Chem.* **450** 289

[13] Uematsu M and Frank E U 1980 Static dielectric constant of water and steam *J. Phys. Chem. Ref. Data* **9** 1291

[14] CRC 2004 *CRC Handbook of Chemistry and Physics* 84th edn (Boca Raton, FL: CRC Press)

[15] http://www.engineeringtoolbox.com/sound-speed-liquids-d_715.html

[16] Williams P R and Williams R L 2004 Cavitation and the tensile strength of liquids under dynamic stressing *Mol. Phys.* **102** 2091

[17] Couzens D C F and Trevena D H 1974 Tensile failure of liquids under dynamic stressing *J. Phys. D: Appl. Phys* **7** 2277

[18] Vaisala 2013 The effect of moisture on the breakdown voltage of transformer oil *Vaisala White Paper* www.vaisala.com/Vaisala%20Documents/White%20Papers/CEN-TIA-power-whitepaper-Moisture-and-Breakdown-Voltage-B211282EN-A-LOW.pdf

[19] Rybak B M 1962 *Analysis of Oil and Oil Products* (Moscow: Gos. Tekh. Izdat.) (in Russian)

[20] Brooks J S and Donnely R J 1977 The calculated thermodynamic properties of superfluid helium-4 *J. Phys. Chem. Ref. Data* **6** 51

[21] Caupin E and Balibar S 2001 Cavitation pressure in liquid helium *Phys. Rev.* B **64** 064507

[22] Peshkov V P and Zinov'eva K N 1959 Experimental work with $^3$He *Rep. Prog. Phys.* **22** 504

[23] Chase C E 1953 Ultrasonic measurements in liquid helium *Proc. R. Soc.* A **220** 116

[24] Heiserman J, Hullin J P, Maynard J and Rudnick I 1976 Precision sound-velocity measurements in He II *Phys. Rev.* B **14** 3862

[25] Atkins K R and Narahara Y 1965 Surface tension of liquid He$^4$ *Phys. Rev.* **138** 437

**IOP** Publishing

Liquid Dielectrics in an Inhomogeneous Pulsed Electric Field

**M N Shneider and M Pekker**

# Chapter 4

## A liquid dielectric in an electric field

*A dielectric as a system of dipoles. Free and bounded charges. Basic electrostatic equations and boundary conditions. The dielectric constant. The energy of the electric field. A dielectric ball in a homogeneous dielectric medium in an external constant electric field. Polarizability of atoms and molecules. Ponderomotive forces in liquid dielectrics. Forces acting on the boundary between two dielectrics.*

### 4.1 A dielectric as a system of dipoles

Dielectrics (or insulators) are substances that do not have freely moving charged particles in an electric field, or their concentration is very low. Dielectrics consist either of neutral molecules (all gaseous and liquid dielectrics, and solid dielectrics) or of positive and negative ions embodied in certain positions of equilibrium (such as crystal lattice nodes). Usually, substances are considered dielectrics if their resistivity is greater than $10^6$ Ohm m. Although the molecules composing dielectrics are neutral, they may have their own dipole moment, or an external electric field can induce a dipole moment on them. Those that have their own permanent dipole moment are called polar molecules (figure 4.1). In non-polar dielectrics, molecules have a symmetrical distribution of positive and negative charges and therefore do not have their own dipole moment (figure 4.2). A typical polar dielectric is ordinary water. Non-polar dielectrics include kerosene, oil, air and inert gases in the liquid and gaseous phases.

In the absence of an external electric field the polar molecules are oriented randomly due to thermal motion, so the time-averaged dipole moment of any volume containing a large number of molecules is zero (figure 4.3(a)). If an external electric field is applied, polar molecules tend to orient along it, making the volume occupied by the insulator acquire a macroscopic dipole moment (figure 4.3(b)). In an external electric field, the electron shells of non-polar molecules are deformed (the

 

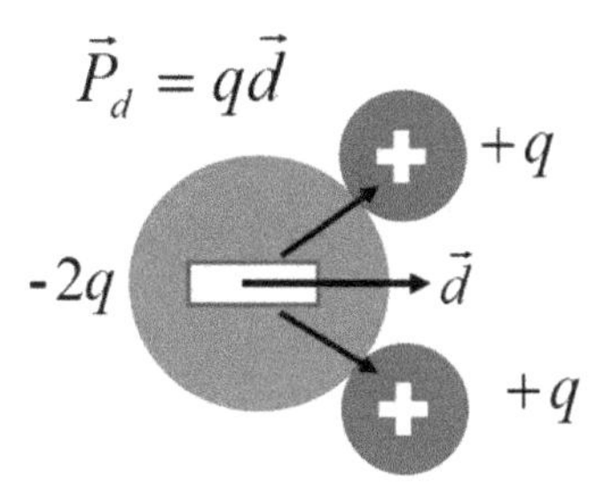


**Figure 4.1.** Schematic view of a polar water molecule $H_2O$.

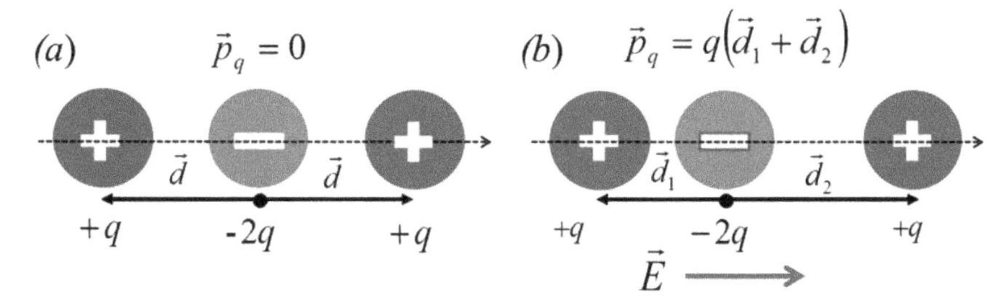


**Figure 4.2.** Schematic view of non-polar molecules, such as $O_2$ or $N_2$. In (a) the dipole moment of the molecule is zero in the absence of an external field and in (b) an external electric field induces a dipole moment on the molecule.

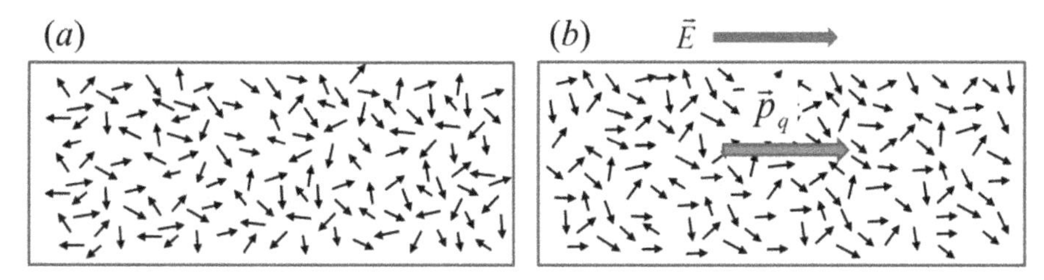


**Figure 4.3.** Polar molecules. (a) The dipoles of the dielectric are randomly oriented in the absence of an external electric field due to thermal motion. As a result, the dipole moment of the dielectric is zero. (b) The dipoles are predominantly oriented along the electric field, and the dipole moment of the dielectric is not zero.

molecules are polarized), just as shown in figure 4.2, so that the volume occupied by the non-polar dielectric also acquires a nonzero dipole moment (figure 4.4(b)). When the electric field is turned off, the total dipole moment vanishes. Note that the uniform electric field does not cause displacement of the molecules in dielectrics (displacement of their center of mass), because they are neutral and the sum of forces acting on them is zero. The electric field only leads to rotation and/or deformation of the molecules.

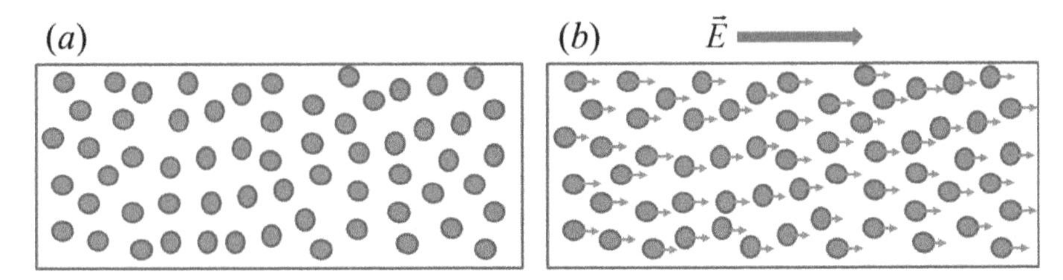


**Figure 4.4.** Non-polar molecules. (a) The dipole moment of each molecule in a dielectric is equal to zero when no external field is applied. (b) The dielectric non-polar molecules acquire induced dipole moments directed along the electric field when an external electric field is applied.

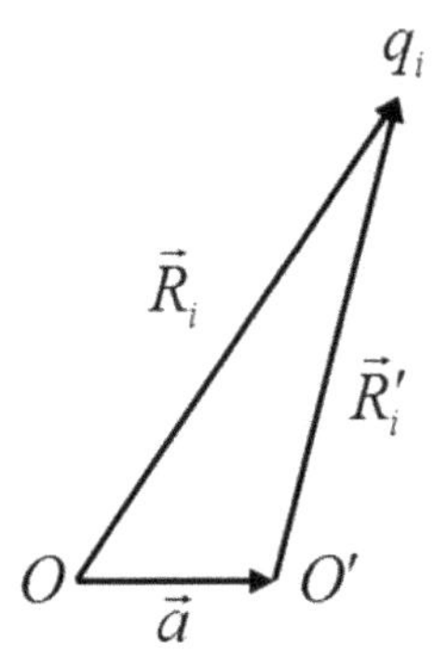


**Figure 4.5.** Offset of the reference point $O$ by vector $\vec{a}$.

We will now find the electric (dipole) moment of the dielectric. By definition, the electric moment of a system of charges $q_i$ is

$$\vec{P}_{\mathrm{d}} = \sum_i q_i \cdot \vec{R}_i. \tag{4.1}$$

Here $\vec{R}_i$ is the radius vector from the selected reference point $O$ to each point charge $q_i$. The sum is taken over all electric charges of the system. The electric moment of the dielectric is independent of the reference point and is equal to the sum of the molecules' dipole moments. Indeed, if the reference point is shifted on the vector $\vec{a}$ from the point $O$ to the point $O'$ (figure 4.5), then

$$\vec{P}'_{\mathrm{d}} = \sum_i q_i \cdot \vec{R}'_i = \vec{p}\,' = \sum_i q_i \cdot \left(\vec{R}_i - \vec{a}\right) = \sum_i q_i \cdot \vec{R}_i - \vec{a}\sum_i q_i = \vec{P}_{\mathrm{d}}, \tag{4.2}$$

in which we took into account that the total charge of the dielectric is zero. The equality of the electric moment of the dielectric to the sum of the dipole moments of molecules follows directly from figure 4.6.

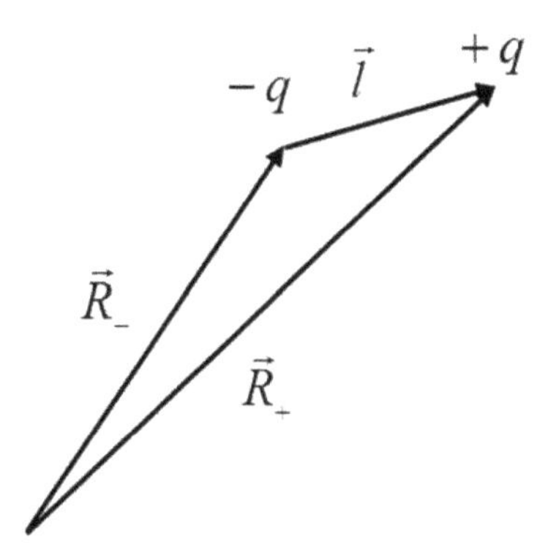


**Figure 4.6.** Electric moment of a system of two charges of opposite sign, equal to the dipole moment $\vec{P}_{\text{d}} = q \cdot \vec{l}$.

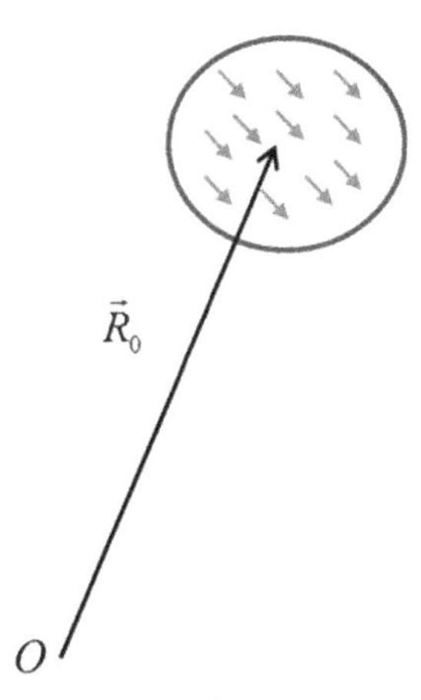


**Figure 4.7.** System of dipoles at a remote distance $\vec{R}_0$ from point $O$, at which the potential is measured.

## 4.2 The potential of a system of dipoles

Now we will calculate the potential created by the system of dipoles at a remote distance $\vec{R}_0$ far from the system (figure 4.7). By the Coulomb law

$$\begin{aligned}\phi_{\text{d}} &= \frac{1}{4\pi\varepsilon_0}\sum_i \frac{q_i}{\left|\vec{R}_0 - \vec{R}_i\right|} = \frac{1}{4\pi\varepsilon_0}\sum_i \frac{q_i}{\left|R_0^2 + R_i^2 - 2\left(\vec{R}_i \cdot \vec{R}_0\right)\right|^{1/2}} \\ &= \frac{1}{4\pi\varepsilon_0 R_0}\sum_i q_i\left(1 + \frac{\left(\vec{R}_i \cdot \vec{R}_0\right)}{R_0^2} - \frac{R_i^2}{R_0^2}\right)q_i \\ &\approx \frac{\vec{R}_0}{4\pi\varepsilon_0 R_0^3} \cdot \sum_i q_i\vec{R}_i = \frac{\left(\vec{R}_0 \cdot \vec{P}_{\text{d}}\right)}{4\pi\varepsilon_0 R_0^3},\end{aligned} \tag{4.3}$$

where $\varepsilon_0 = 8.8542 \cdot 10^{-12}$ [F m$^{-1}$] is the dielectric constant of a vacuum. The approximation at the end takes into account that the point at which the potential is calculated is remote from the system of charges at a distance much larger than the size of the system (figure 4.7). It follows from (4.3) that the neutral system of charges with the electric moment $\vec{P}_{\rm d}$ is equivalent to a single dipole with the dipole moment $\vec{P}_{\rm d}$.

In integrated form (4.3) becomes

$$\phi_{\rm d} = \frac{1}{4\pi\varepsilon_0}\int_V \frac{\left(\vec{R}_0 - \vec{r}\right)\cdot \vec{p}_{\rm d}(\vec{r})}{\left|\vec{R}_0 - \vec{r}\right|^3}{\rm d}V. \tag{4.4}$$

Here, $\vec{p}_{\rm d}(\vec{r}) = \frac{1}{\Delta V}\sum_{i\subset \Delta V}\vec{P}_{{\rm d},i}$ is the average amount of dipole moment in the volume $\Delta V$. Evidently, the size of the averaging volume $\Delta V$ should be sufficiently large so that the number of molecules in it is much greater than one. Since

$$\frac{\left(\vec{R}_0 - \vec{r}\right)}{\left|\vec{R}_0 - \vec{r}\right|^3} = \vec{\nabla}_{\vec{r}}\left(\frac{1}{\left|\vec{R}_0 - \vec{r}\right|}\right), \tag{4.5}$$

according to the formulas of vector analysis, equation (4.4) can be rewritten as

$$\begin{aligned}\phi_{\rm d} &= \frac{1}{4\pi\varepsilon_0}\int_V \frac{(\vec{R}_0 - \vec{r})\cdot \vec{p}_{\rm d}(\vec{r})}{\left|\vec{R}_0 - \vec{r}\right|^3}{\rm d}V = \frac{1}{4\pi\varepsilon_0}\int_V \vec{\nabla}\cdot\left(\frac{\vec{p}_{\rm d}(\vec{r})}{\left|\vec{R}_0 - \vec{r}\right|}\right){\rm d}V \\ &- \frac{1}{4\pi\varepsilon_0}\int_V \frac{\vec{\nabla}\cdot\vec{p}_{\rm d}(\vec{r})}{\left|\vec{R}_0 - \vec{r}\right|}{\rm d}V.\end{aligned} \tag{4.6}$$

According to the Gauss–Ostrogradsky theorem (divergence theorem), the first integral over the volume on the right-hand side of (4.6) can be rewritten as an integral over the surface,

$$\frac{1}{4\pi\varepsilon_0}\int_V \vec{\nabla}\cdot\left(\frac{\vec{p}_{\rm d}(\vec{r})}{\left|\vec{R}_0 - \vec{r}\right|}\right){\rm d}V = \frac{1}{4\pi\varepsilon_0}\int_S \frac{p_{\rm d,n}(\vec{r})}{\left|\vec{R}_0 - \vec{r}\right|}{\rm d}s + \frac{1}{4\pi\varepsilon_0}\int_{S_1} \frac{p_{\rm d,n}(\vec{r})}{\left|\vec{R}_0 - \vec{r}\right|}{\rm d}s, \tag{4.7}$$

where $S$ is the external surface area of the considered volume $V$, and $S_1$ are the corresponding areas of all surfaces where the polarization vector has a discontinuity (figure 4.8).

If we consider the total field, that is we put the boundary $S$ of the computational domain at infinity, then the first integral on the right-hand side of (4.7) equals zero. In turn, the integral over the surfaces is

$$\frac{1}{4\pi\varepsilon_0}\int_{S_1} \frac{p_{\rm d,n}(\vec{r})}{\left|\vec{R}_0 - \vec{r}\right|}{\rm d}S = \frac{1}{4\pi\varepsilon_0}\int_{S_1} \frac{p_{{\rm d}_1,{\rm n}} - p_{{\rm d}_2,{\rm n}}}{\left|\vec{R}_0 - \vec{r}\right|}{\rm d}S, \tag{4.8}$$

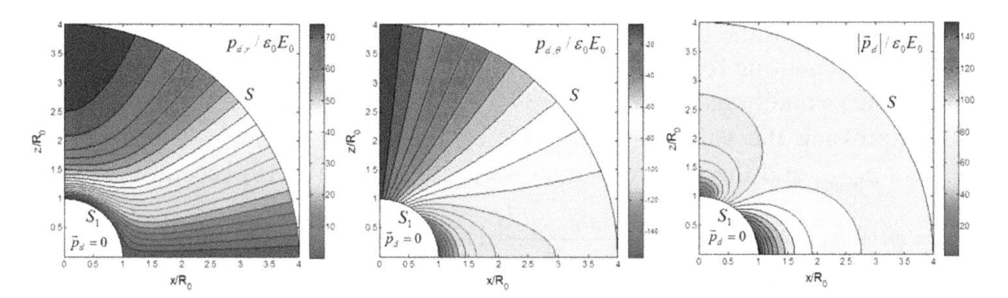


**Figure 4.8.** A spherical bubble of air in water ($\varepsilon = 81$) in a constant homogeneous electric field $E_0$ directed along the $z$-axis.

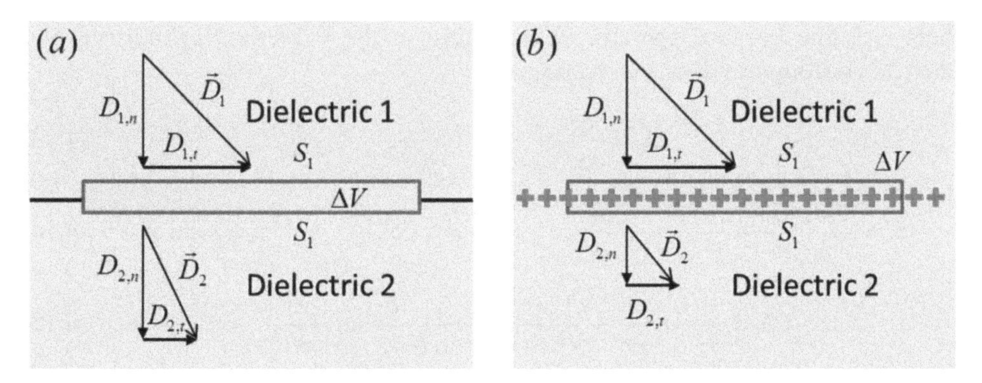


**Figure 4.9.** The interface between two dielectrics. In (a) there are no free surface charges on the boundary. The displacements in dielectrics 1 and 2 are different, but the normal displacement components are equal. In (b) there are free surface charges on the border, so the normal component of the displacement is discontinuous $D_{2,\mathrm{n}} - D_{1,\mathrm{n}} = \sigma_\mathrm{f}$.

where $p_{\mathrm{d_1,n}}$ and $p_{\mathrm{d_2,n}}$ are the normal components of the dipole moment inside and outside region $S_1$, respectively (figure 4.9).

Using formulas (4.7) and (4.8), it is convenient to rewrite the expression for $\phi_\mathrm{d}$ in (4.6)

$$\phi_\mathrm{d} = -\frac{1}{4\pi\varepsilon_0}\int_V \frac{\vec{\nabla}\cdot\vec{p}_\mathrm{d}(\vec{r})}{\left|\vec{R}_0 - \vec{r}\right|}\mathrm{d}V - \frac{1}{4\pi\varepsilon_0}\int_{S_1} \frac{p_{\mathrm{d_2,n}} - p_{\mathrm{d_1,n}}}{\left|\vec{R}_0 - \vec{r}\right|}\mathrm{d}s. \tag{4.9}$$

In general, the potential at point $\vec{R}_0$ consists of the potential of free charges $\phi_\mathrm{f}$ and the potential of bounded charges of the dipoles of the dielectric $\phi_\mathrm{d}$,

$$\begin{aligned}\phi = \phi_\mathrm{f} + \phi_\mathrm{d} = &\frac{1}{4\pi\varepsilon_0}\int_V \frac{\rho_\mathrm{f}(\vec{r})}{\left|\vec{R}_0 - \vec{r}\right|}\mathrm{d}V + \frac{1}{4\pi\varepsilon_0}\int_V \frac{\sigma_\mathrm{f}(\vec{r})}{\left|\vec{R}_0 - \vec{r}\right|}\mathrm{d}s \\ &- \frac{1}{4\pi\varepsilon_0}\int_V \frac{\vec{\nabla}\cdot\vec{p}_\mathrm{d}(\vec{r})}{\left|\vec{R}_0 - \vec{r}\right|}\mathrm{d}V - \frac{1}{4\pi\varepsilon_0}\int_{S_1} \frac{p_{\mathrm{d_2,n}} - p_{\mathrm{d_1,n}}}{\left|\vec{R}_0 - \vec{r}\right|}\mathrm{d}s\end{aligned}, \tag{4.10}$$

where $\rho_{\rm f}$ and $\sigma_{\rm f}$ are the volume and surface densities of free charges. We will show below that accounting for surface discontinuities can be reduced to the corresponding boundary conditions.

By expressing the values $-\vec{\nabla}\cdot\vec{p}_{\rm d}(\vec{r})$ and $p_{{\rm d}_2,{\rm n}} - p_{{\rm d}_1,{\rm n}}$ by the bounded volume charges $-\rho_{\rm bound}$ and the surface charges $-\sigma_{\rm bound}$, equation (4.10) becomes

$$\phi = \phi_{\rm f} + \phi_{\rm d} = \frac{1}{4\pi\varepsilon_0}\int_V \frac{\rho_{\rm f}(\vec{r}) + \rho_{\rm bound}(\vec{r})}{\left|\vec{R}_0 - \vec{r}\right|}{\rm d}V + \frac{1}{4\pi\varepsilon_0}\int_{S_1} \frac{\sigma_{\rm f}(\vec{r}) + \sigma_{\rm bound}(\vec{r})}{\left|\vec{R}_0 - \vec{r}\right|}{\rm d}s. \quad (4.11)$$

Taking into account that $\vec{E} = -\vec{\nabla}\varphi$, we obtain

$$\vec{\nabla}\cdot\vec{E} = -\Delta\phi = \frac{1}{\varepsilon_0}\left(\rho_{\rm f}(\vec{r}) - \vec{\nabla}\cdot\vec{p}_{\rm d}(\vec{r})\right), \quad (4.12)$$

where $\Delta$ is the Laplace operator, whose form in the Cartesian, cylindrical, and spherical coordinate systems is, respectively,

$$\Delta = \frac{\partial^2}{\partial x^2} + \frac{\partial^2}{\partial y^2} + \frac{\partial^2}{\partial z^2}, \quad (4.13)$$

$$\Delta = \frac{1}{r}\frac{\partial}{\partial r}r\frac{\partial}{\partial r} + \frac{1}{r^2}\frac{\partial^2}{\partial\varphi^2} + \frac{\partial^2}{\partial z^2}, \qquad x = r\cos\varphi, \qquad y = r\sin\varphi, \quad (4.14)$$

$$\Delta = \frac{1}{r^2}\frac{\partial}{\partial}r^2\frac{\partial}{\partial r} + \frac{1}{r^2}\frac{\partial^2}{\partial\varphi^2} + \frac{1}{\sin\theta}\frac{\partial}{\partial\theta}\sin\theta\frac{\partial}{\partial\theta}, \quad (4.15)$$

$x = r\sin\theta\cos\phi$, $y = r\sin\theta\sin\phi$, $z = r\cos\theta$.

In the transition from expression (4.11) to (4.12), we took into account that

$$\Delta\frac{1}{\left|\vec{R}_0 - \vec{r}\right|} = \frac{2}{\left|\vec{R}_0 - \vec{r}\right|^2}\delta\left|\vec{R}_0 - \vec{r}\right|, \quad (4.16)$$

in which $\delta|\vec{R}_0 - \vec{r}|$ is the Dirac delta-function ($\delta(\xi) = 0$ for $|\xi| > 0$ and $\int_{-\infty}^{\infty}\delta(\xi){\rm d}\xi = 1$) and

$$\int_V \Delta_{\vec{R}_0}\frac{\Psi(\vec{r})}{\left|\vec{R}_0 - \vec{r}\right|}{\rm d}V = \int_{|\vec{R}_0-\vec{r}|}^{\infty}\Psi(\vec{r})\frac{2\delta\left(\left|\vec{R}_0 - \vec{r}\right|\right)}{\left|\vec{R}_0 - \vec{r}\right|^2}\left|\vec{R}_0 - \vec{r}\right|^2 \times \sin\theta{\rm d}\theta\cdot{\rm d}\varphi\cdot{\rm d}\left|\vec{R}_0 - \vec{r}\right| = 4\pi\Psi\left(\vec{R}_0\right), \quad (4.17)$$

where $\Psi(\vec{r})$ is an arbitrary function of a vector argument.

Introducing the electric displacement vector

$$\vec{D} = \varepsilon_0\vec{E} + \vec{p}_{\rm d}, \quad (4.18)$$

the Poisson equation follows from (4.12) and (4.18):

$$\vec{\nabla}\cdot\vec{D} = \rho_{\rm f}(\vec{r}). \quad (4.19)$$

Given equation (4.18), the normal component of vector $\vec{D}$ is continuous at the boundary. Indeed, consider the interface surface (figure 4.8). Since the volume integral over the volume $\Delta V$ can be transformed into an integral over the surface $S_1$, bounding this volume yields

$$\int_{\Delta V} \left(\vec{\nabla} \cdot \vec{D}\right) \mathrm{dV} = \int_{S_2} D_{2,\mathrm{n}} \mathrm{d}S_2 - \int_{S_1} D_{1,\mathrm{n}} \mathrm{d}S_1 = 0, \quad (4.20)$$

or

$$D_{2,\mathrm{n}} = D_{1,\mathrm{n}}. \quad (4.21)$$

In the integral of (4.20), we neglected the integrals over the side surfaces of the parallelepiped surface of integration (figure 4.9), because their area can be made arbitrarily small.

If, however, the interface has a nonzero surface charge density $\sigma_{\mathrm{f}}$, the normal component of the displacement vector is not continuous (figure 4.9(b)):

$$D_{2,\mathrm{n}} - D_{1,\mathrm{n}} = \sigma_{\mathrm{f}}. \quad (4.22)$$

Now we will show that the tangential component of the electric field is continuous on the interface surface. Due to the potentiality of the electric field vector in electrostatics, the circulation of a vector $\vec{E}$ along an arbitrary closed contour near the interface of a dielectric is zero (figure 4.10). Therefore,

$$\oint \vec{E} \cdot \mathrm{d}\vec{l} = E_{1,\mathrm{t}} l_1 - E_{2,\mathrm{t}} l_1 = 0 \quad (4.23)$$

and

$$E_{1,\mathrm{t}} = E_{2,\mathrm{t}}. \quad (4.24)$$

In the integral of (4.23), we neglected the integrals over the sides of the integration path along the normals to the interface surface (figure 4.10), because their length can be made arbitrarily small.

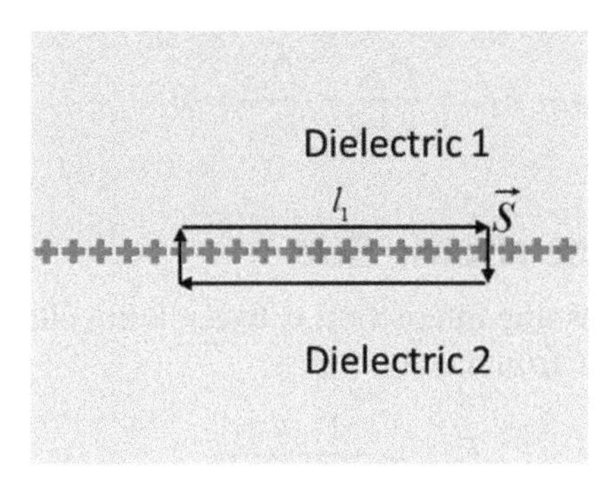


**Figure 4.10.** The integration contour in the vicinity of the charged interface surface between two dielectrics.

## 4.3 The dielectric constant

The dipole moment of the volume occupied by the dielectric is absent at zero external electric field, regardless of what kind of molecules compose it: polar or non-polar. Therefore, it is natural to expect that, in a sufficiently weak field, the dipole moment per unit volume of a dielectric is proportional to the electric field:

$$\vec{p}_{\rm d} = \varepsilon_0 \chi \vec{E}. \tag{4.25}$$

The proportionality factor $\chi$ in (4.25) is called the dielectric susceptibility. In general, it may depend on the direction of the electric field, and is a tensor:

$$p_{q,i} = \varepsilon_0 \chi_{ik} E_k. \tag{4.26}$$

The tensor dielectric susceptibility is characteristic for solid dielectrics having a crystalline structure, and for liquid crystals. For isotropic liquids and gaseous dielectrics, the dielectric susceptibility is scalar.

At very large fields, the dipole moment acquired by a dielectric cannot exceed the sum of the dipole moments of all its molecules. However, such fields cannot be achieved in practice, because the processes of molecular dissociation and ionization begin earlier and, as a result of dielectric breakdown, the dielectric becomes a conductor.

Substituting (4.25) into (4.18), we obtain the displacement vector

$$\vec{D} = \varepsilon_0(1 + \chi)\vec{E} = \varepsilon_0 \varepsilon \vec{E}, \tag{4.27}$$

where $\varepsilon = (1 + \chi)$ is the relative permittivity of the isotropic medium. In general, the dielectric constant $\varepsilon$ depends on the composition of the medium, the thermodynamic parameters of temperature and density, and the frequency and value of the electromagnetic field. We give typical values for the dielectric permittivity of liquid dielectrics in chapter 3.

For convenience, we summarize the basic formulas derived above,

$$\vec{E} = -\vec{\nabla}\phi, \qquad \vec{D} = \varepsilon_0 \varepsilon \vec{E}, \qquad \vec{p}_{\rm d} = \varepsilon_0(\varepsilon - 1)\vec{E}, \qquad \vec{\nabla} \cdot \vec{D} = \rho_{\rm f}, \tag{4.28}$$

along with the boundary conditions at the interface between two dielectrics:

$$\begin{aligned} D_{2,\rm n} - D_{1,\rm n} &= \varepsilon_0 \varepsilon_2 E_{2,\rm n} - \varepsilon_0 \varepsilon_2 E_{1,\rm n} = \sigma_{\rm f} \\ E_{2,\rm t} - E_{1,\rm t} &= \frac{D_{2,\rm t}}{\varepsilon_0 \varepsilon_2} - \frac{D_{1,\rm t}}{\varepsilon_0 \varepsilon_1} = 0. \end{aligned} \tag{4.29}$$

## 4.4 The energy of the electric field

The energy of the electrostatic interaction between point charges $q_1$ and $q_2$ located in a vacuum at a distance $r$ from each other is

$$W_{1,2} = \frac{1}{4\pi\varepsilon_0}\frac{q_1 q_2}{r}. \tag{4.30}$$

If the charges are of the same sign they repel each other and energy (4.30) equals the kinetic energy of these charges at infinity. If the charges are of opposite signs, the work necessary to separate these charges to infinity equals (4.30) with the opposite sign.

Formula (4.30) can be written as

$$W_{1,2} = \frac{1}{2}\left(\frac{1}{4\pi\varepsilon_0}\frac{q_1}{r}q_2 + \frac{1}{4\pi\varepsilon_0}\frac{q_2}{r}q_1\right) = \frac{1}{2}(\varphi_1 q_2 + \varphi_2 q_1), \tag{4.31}$$

where $\varphi_1$ is the potential that creates a second charge at the point where the first charge is located, and $\varphi_2$ is likewise the potential of the first charge at the position of the second charge. Accordingly, the electrostatic energy of interaction between $N$ charges is

$$W = \frac{1}{2}\sum_{i\neq j}^{N}\frac{1}{4\pi\varepsilon_0}\frac{q_i q_j}{r_{ij}} = \frac{1}{2}\sum_{i=1}^{N}\varphi_i q_i, \tag{4.32}$$

in which $r_{ij}$ is the distance between the $i$th and $j$th charges, and $\varphi_i$ is the potential that creates all the charges except the $i$th one at its location. The interaction of the charges is not considered, because it goes to infinity when the radii of the point charges tend to zero.

We now generalize this expression for the potential energy of electric charges to an arbitrary case, including dielectric media. Assume there is a charge $\mathrm{d}q = \rho_\mathrm{f}\mathrm{d}V$ in a volume element $\mathrm{d}V$ in the discrete distribution of the potential created by all the free and bound charges of the system. In this case, the total potential energy of all free charges in volume $V$ is

$$W = \frac{1}{2}\int_V \phi\rho_\mathrm{f}\mathrm{d}V. \tag{4.33}$$

The transition from (4.32) to (4.33) implies that volume $\mathrm{d}V$ contains enough point charges so that the charge density $\rho_\mathrm{f} = \frac{1}{\mathrm{d}V}\sum_{i\subset \mathrm{d}V} q_i$ can be considered to be a continuous function of the coordinates. In other words the charge 'is spread' in space and as $\mathrm{d}V \to 0$, $\mathrm{d}q = \rho_\mathrm{f}\mathrm{d}V \to 0$.

By substituting formula (4.19) for $\rho_\mathrm{f}$ into equation (4.33), we obtain

$$W = \frac{1}{2}\int_V \phi\left(\vec{\nabla}\cdot\vec{D}\right)\mathrm{d}V = \frac{1}{2}\int_V\left(\vec{D}\cdot\vec{E}\right)\mathrm{d}V + \frac{1}{2}\int_V \vec{\nabla}\cdot\left(\phi\vec{D}\right)\mathrm{d}V, \tag{4.34}$$

in which we used the formula of vector analysis $\varphi(\vec{\nabla}\cdot\vec{D}) = -\vec{D}\cdot\vec{\nabla}\varphi + \vec{\nabla}(\varphi\cdot\vec{D})$ and $\vec{E} = -\vec{\nabla}\varphi$. The second integral in the right-hand side of equation (4.34) can be conveniently transformed using the Gauss–Ostrogradsky theorem to

$$\int_V \vec{\nabla}\cdot\left(\phi\vec{D}\right)\mathrm{d}V = \int_{\vec{S}}\phi\vec{D}\mathrm{d}\vec{S}, \tag{4.35}$$

where $S$ is a closed surface surrounding volume $V$. From the assumption that all bound and free charges are within a finite region, we find that when increasing the distance between them $\varphi \propto 1/r$ and $|\vec{D}| \propto 1/r^2$. That is, $\varphi|\vec{D}| \propto 1/r^3$. Since the area $S$ grows as $r^2$, the integral $\int_S \phi\vec{D}\mathrm{d}\vec{S} \propto 1/r$, and it tends to zero when the surface of integration moves to infinity. Therefore, for the integral over the entire volume,

$$W = \frac{1}{2}\int\left(\vec{D}\cdot\vec{E}\right)\mathrm{d}V = \frac{\varepsilon_0}{2}\int \varepsilon E^2 \mathrm{d}V. \tag{4.36}$$

The integral in (4.33) is taken only over the region of space where there are free charges ($\rho_f \neq 0$), but in the form (4.36) it is expressed in terms of the electric field, and bonded charges (dipoles) are included implicitly via the dielectric constant.

The energy density in (4.36) is equal to

$$w = \frac{1}{2}\vec{E}\cdot\vec{D} = \frac{1}{2}\varepsilon\varepsilon_0 E^2 \geqslant 0. \tag{4.37}$$

Since expression (4.37) for the energy density is formulated in the local form and determines the energy density as a function of the electric field and the dielectric properties of the medium at a given point, the validity of (4.36) cannot depend on the way in which a field at a given point of space was created. Therefore, the expression for the energy is not only true for the permanent electrostatic fields, but also for the alternating fields.

To calculate the electrical field in (4.37) it is possible to use formula (4.10), which allows one to find the electric field at any spatial point. Formula (4.10) takes into account not only the free and bound volume charges, but also the surface charges.

Equation (4.33) can be generalized to the case when the medium contains not only space charges, but also surface charges:

$$W = \frac{1}{2}\int_V \phi\rho_q \mathrm{d}V + \frac{1}{2}\int_{S_1} \phi\sigma_1 \mathrm{d}S_1. \tag{4.38}$$

The second integral on the right-hand side of (4.38) is carried out over all surfaces where there is surface charge density.

## 4.5 Energy of a dielectric in an external electric field

For simplicity, we assume that the dielectric occupies the entire space. Suppose that an electrostatic field is generated by a system of free charges. Then, in accordance with (4.36), the energy of the field in the absence of a dielectric is $W_0 = \frac{1}{2}\varepsilon_0\int E_0^2 \mathrm{d}V$, and in the presence of a dielectric is $W = \frac{1}{2}\varepsilon_0\int \varepsilon E^2 \mathrm{d}V$, wherein $\varepsilon E = E_0$. Accordingly, the electrostatic energy acquired by a dielectric in an external field is

$$\begin{aligned} W_{\text{diel}} = W - W_0 &= \frac{1}{2}\varepsilon_0\int\left(\varepsilon E^2 - E_0^2\right)\mathrm{d}V = -\frac{1}{2}\varepsilon_0\int(\varepsilon - 1)\left(\vec{E}\cdot\vec{E}_0\right)\mathrm{d}V \\ &= -\frac{1}{2}\int\left(\vec{p}_\text{d}\cdot\vec{E}_0\right)\mathrm{d}V. \end{aligned} \tag{4.39}$$

This equation also remains valid in the case of a dielectric of finite size.

## 4.6 A dielectric ball in a homogeneous dielectric medium in an external constant electric field

We will now consider the classical problem of a dielectric ball of radius $R$ in a constant electric field $E_0$ directed along the $z$-axis. The dielectric constants of the ball and the surrounding media are $\varepsilon_{in}$ and $\varepsilon_{out}$, respectively.

The potential of the electric field outside the ball has the form

$$\phi_{out} = -E_0 r \cos(\theta) + A \cdot E_0 \frac{\cos(\theta)}{r^2}. \tag{4.40}$$

The first term on the right-hand side of this equation corresponds to the external field, and the second term to the influence of the dielectric ball. The potential inside the ball is

$$\phi_{in} = -E_0 B r \cos(\theta). \tag{4.41}$$

The conditions for the potentials on the boundary of the ball follow from the boundary conditions (4.29):

$$\begin{cases} \left.\frac{1}{R}\frac{\partial\phi_{out}}{\partial\theta}\right|_R = \left.\frac{1}{R}\frac{\partial\phi_{in}}{\partial\theta}\right|_R, & -1 + \frac{A}{R^3} = -B \\ \left.\varepsilon_{out}\frac{\partial\phi_{out}}{\partial r}\right|_R = \left.\varepsilon_{in}\frac{\partial\phi_{in}}{\partial r}\right|_R, & -\varepsilon_{out}\left(1 + \frac{2A}{R^3}\right) = -\varepsilon_{in} B \end{cases} \tag{4.42}$$

yielding

$$A = \frac{\varepsilon_{in} - \varepsilon_{out}}{\varepsilon_{in} + 2\varepsilon_{out}} R^3, \qquad B = \frac{3\varepsilon_{out}}{\varepsilon_{in} + 2\varepsilon_{out}}. \tag{4.43}$$

Accordingly, from (4.40) and (4.41) we obtain

$$\phi_{out} = -E_0 r\left(1 + \frac{(\varepsilon_{out} - \varepsilon_{in})}{2\varepsilon_{out} + \varepsilon_{in}}\frac{R^3}{r^3}\right)\cos(\theta), \quad \phi_{in} = -E_0 r \cos(\theta)\left(\frac{3\varepsilon_{out}}{2\varepsilon_{out} + \varepsilon_{in}}\right) \tag{4.44}$$

$$\begin{aligned} E_{out,r} &= E_0\left(1 - 2\frac{(\varepsilon_{out} - \varepsilon_{in})}{2\varepsilon_{out} + \varepsilon_{in}}\frac{R^3}{r^3}\right)\cos(\theta), \\ E_{out,\theta} &= -E_0\left(1 + \frac{(\varepsilon_{out} - \varepsilon_{in})}{2\varepsilon_{out} + \varepsilon_{in}}\frac{R^3}{r^3}\right)\sin(\theta) \end{aligned} \tag{4.45}$$

$$E_{in,r} = \frac{3\varepsilon_{out}}{\varepsilon_{in} + 2\varepsilon_{out}} E_0 \cos(\theta), \qquad E_{in,\theta} = -\frac{3\varepsilon_{out}}{\varepsilon_{in} + 2\varepsilon_{out}} E_0 \sin(\theta) \tag{4.46}$$

$$D_{out,r} = \varepsilon_{out} E_{out,r}, \qquad D_{out,\theta} = \varepsilon_{out} E_{out,\theta} \tag{4.47}$$

$$D_{in,r} = \varepsilon_{in} E_{in,r}, \qquad D_{in,\theta} = \varepsilon_{in} E_{in,\theta}. \tag{4.48}$$

The potential $\phi_{\text{out}}$ in (4.44) is composed of the unperturbed electric field potential $-\vec{E}_0 \cdot \vec{r}$ and

$$\delta\phi_{\text{out}} = -\frac{(\varepsilon_{\text{out}} - \varepsilon_{\text{in}})}{2\varepsilon_{\text{out}} + \varepsilon_{\text{in}}} R^3 \frac{\left(\vec{E}_0 \cdot \vec{r}\right)}{r^3}, \tag{4.49}$$

which can be interpreted as the induced dipole potential. Indeed, according to (4.3), the potential of the dipole $\vec{P}_{\text{d}}$ at the point $\vec{r}$ is

$$\phi_{\text{d}} = \frac{1}{4\pi\varepsilon_0} \frac{\vec{P}_{\text{d}} \cdot \vec{r}}{r^3}. \tag{4.50}$$

Comparing (4.49) with (4.50), we obtain

$$\vec{P}_{\text{d,ball}} = -4\pi\varepsilon_0 R^3 \frac{(\varepsilon_{\text{out}} - \varepsilon_{\text{in}})}{2\varepsilon_{\text{out}} + \varepsilon_{\text{in}}} \vec{E}_0. \tag{4.51}$$

Accordingly, the volume density of the dipole moment of the ball is equal to

$$\vec{p}_{\text{d,ball}} = \frac{\vec{P}_{\text{d,ball}}}{V_{\text{ball}}} = -\frac{1}{4\pi R^3/3} 4\pi\varepsilon_0 \vec{E}_0 \frac{(\varepsilon_{\text{out}} - \varepsilon_{\text{in}})}{2\varepsilon_{\text{out}} + \varepsilon_{\text{in}}} R^3 = 3\varepsilon_0 \frac{(\varepsilon_{\text{in}} - \varepsilon_{\text{out}})}{2\varepsilon_{\text{out}} + \varepsilon_{\text{in}}} \vec{E}_0 \tag{4.52}$$

$$E_{\text{in}} = E_0 - \frac{p_{\text{d,ball}}}{3\varepsilon_0} = \frac{3\varepsilon_{\text{out}}}{\varepsilon_{\text{in}} + 2\varepsilon_{\text{out}}} E_0. \tag{4.53}$$

For a spherical vapor or vacuum bubble in a dielectric, $\varepsilon_{\text{in}} = 1$, $\varepsilon_{\text{out}} = \varepsilon$ and $\vec{p}_{\text{d,ball}} = -3\varepsilon_0 \frac{(\varepsilon - 1)}{2\varepsilon + 1} \vec{E}_0$ . That is, the dipole moment of the bubble is directed against the external electric field.

Figure 4.11 shows the two-dimensional distribution of the amplitude of the electric field, calculated from (4.45) and (4.46), for three cases: a spherical bubble of air in water ($\varepsilon_{\text{out}} = 81$, $\varepsilon_{\text{in}} = 1$) (a), a spherical bubble of air in transformer oil

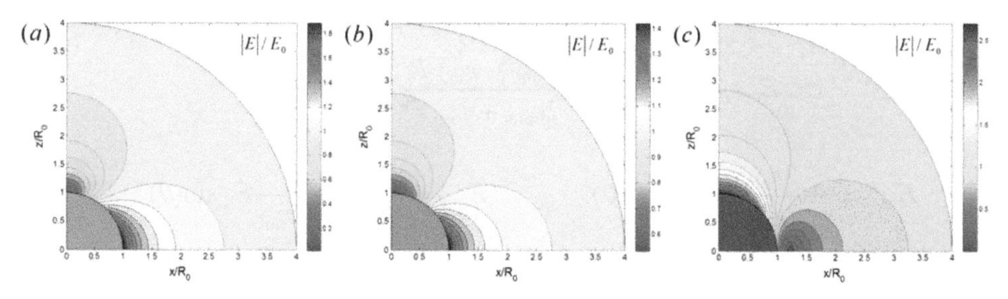


**Figure 4.11.** Two-dimensional distributions of the normalized amplitude of the electric field for three cases in a static electric field $E_0$ directed along the $z$-axis: a spherical bubble of air in water (a), a spherical bubble of air in transformer oil (b), and a spherical droplet of water in transformer oil (c).

($\varepsilon_{out} = 2.3$, $\varepsilon_{in} = 1$) (b), and a spherical droplet of water in transformer oil ($\varepsilon_{out} = 2.3$, $\varepsilon_{in} = 81$) (c). In the first two cases, the field inside the bubble exceeds the field at infinity, but the field inside the water droplet in transformer oil is noticeably less than the field at infinity, and the field at the pole just outside the water droplet is more than 2.5 times greater than the field at infinity. Thus, droplets of water in insulating (transformer) oil can trigger breakdown of the insulation in high-voltage devices (for more details see chapters 5 and 6, and [1]).

## 4.7 Polarizability of atoms and molecules

All atoms and molecules in an electric field acquire an induced dipole moment $\vec{P}_d$, which is a linear function of the electric field and is directed along the field,

$$\vec{P}_d = \alpha_e \vec{E}, \tag{4.54}$$

where coefficient $\alpha_e$ is the electronic polarizability of the atom or molecules.

### 4.7.1 Non-polar dielectrics (the Clausius–Mossotti relation)

We will estimate the polarizability of atoms or non-polar molecules using a simple example. Assume that a non-polar molecule or atom is a uniformly charged sphere of radius $R_{ball}$ and charge $-q$ that is centered on a point nucleus of charge $+q$ (figure 4.12(a)). The dipole moment of this molecule (atom) is zero.

It is easy to find the electric field of the negatively charged sphere acting on the nucleus from the Poisson equation written in spherical coordinates,

$$\frac{1}{r^2}\frac{\partial}{\partial r}r^2\frac{\partial \phi}{\partial r} = -\frac{1}{r^2}\frac{\partial}{\partial r}r^2 E_r = -\frac{3q}{4\pi\varepsilon_0 R_{ball}^3} \tag{4.55}$$

$$\vec{E}_r(r) = \frac{q\vec{r}}{4\pi\varepsilon_0 R_{ball}^3}, \tag{4.56}$$

where $\vec{r}$ is the coordinate of the position of a point nucleus relative to the center (figure 4.12). Under the influence of an external field, the center of the negatively

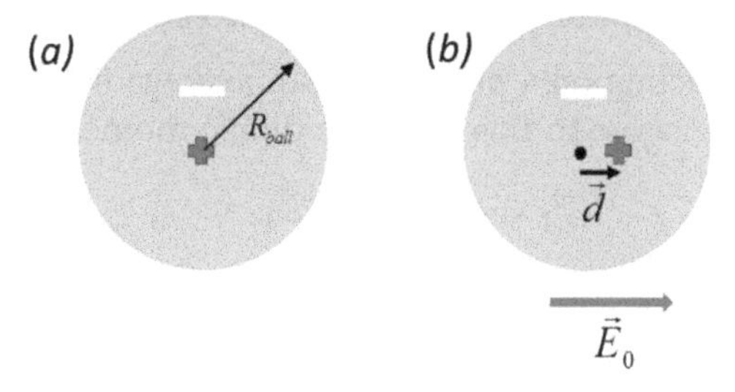


**Figure 4.12.** A schematic view of a non-polar molecule. In the absence of an external electric field, the dipole moment of the molecule is zero (a). But under the influence of an external electric field, the center of the electron shell of the molecule (atom) moves from the positively charged point nucleus to a distance such that the sum of forces acting on the nucleus by the external field and the negatively charged sphere compensate each other (b).

**Table 4.1.** Static average electronic dipole polarizabilities for the ground states of some atoms and molecules (in units $10^{-30}$ [m$^3$]) [2].

| Atom | Polarizability | Molecule | Polarizability |
|---|---|---|---|
| He | 0.205 | $N_2$ | 1.74 |
| C | 1.76 | $O_2$ | 1.58 |
| N | 1.10 | $H_2O$ | 1.45 |
| O | 0.802 | $CO_2$ | 2.91 |
| Ar | 1.641 | $N_2O$ | 3.03 |
| Xe | 4.044 | $NO_2$ | 3.02 |

charged sphere shifts with respect to the positive point nucleus to a distance $d$ such that the restoring force by the negative charge compensates the influence of the external field. If the value of the external electric field is $E_0$, then from (4.56) the shift is equal to

$$d = \frac{4\pi\varepsilon_0 R_{\text{ball}}^3 E_0}{q}. \tag{4.57}$$

Accordingly, the induced dipole moment of the molecule (atom) is equal to

$$\vec{P}_{\text{d,e}} = q\vec{d} = 4\pi\varepsilon_0 R_{\text{ball}}^3 \vec{E}_0. \tag{4.58}$$

Then by (4.54), the electronic polarizability $\alpha_e$ is

$$\alpha_e = 4\pi\varepsilon_0 R_{\text{ball}}^3. \tag{4.59}$$

In general, $\alpha_e$ differs from $4\pi\varepsilon_0 R_{\text{ball}}^3$, because the density of the electron cloud varies radially and has no precise boundaries. In SI units, polarizability is measured in C m$^2$ V$^{-1}$, or is represented by $\alpha_e = 4\pi\varepsilon_0\alpha_e'$, in which $\alpha_e'$ is measured in units of volume m$^{-3}$. Table 4.1 lists the $\alpha'_e$ coefficients of various atoms and molecules.

### 4.7.2 The dielectric constant of dense non-polar dielectric media

Assume that the field acting on a molecule is an average field in a substance that is valid only for rarefied gases. In this case, the dipole moment per unit volume is

$$\vec{P}_{\text{d}} = n\vec{P}_{\text{d,e}} = n\alpha_e\vec{E}_0. \tag{4.60}$$

Here $n$ is the concentration of non-polar molecules in the gas.

For a gas, the influence of the electric fields of neighboring atoms and molecules can be neglected, because the distances between the gas molecules are much greater than the size of the dipole. In liquids and solids, however, it cannot be neglected. We will apply the following standard method to find the average field acting on the molecule in a liquid or solid. We will assume that each non-polar molecule is inside a kind of 'spherical hole' (figure 4.13).

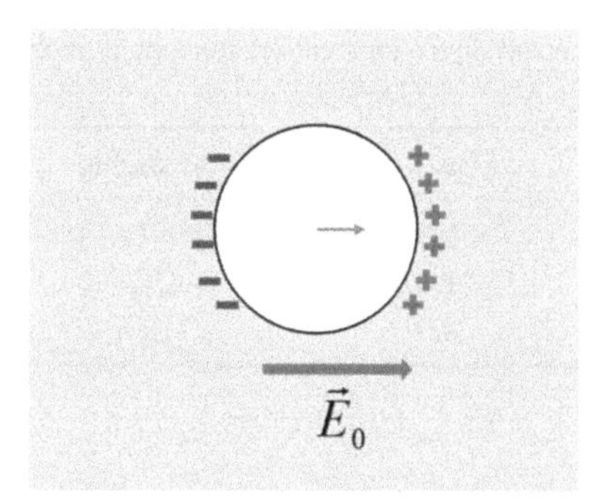


**Figure 4.13.** A non-polar molecule inside a dielectric. The blue arrow indicates the induced displacement of the electron cloud relative to the nucleus.

In the case of a bubble in a dielectric with $\varepsilon_{\text{in}} = 1$ and $\varepsilon_{\text{out}} = \varepsilon$, the density of the induced dipole moment is

$$\vec{p}_{\text{d,B}} = -3\varepsilon_0 \frac{\varepsilon - 1}{2\varepsilon + 1} \vec{E}_0. \tag{4.61}$$

Then, the expression for the electric displacement is

$$\begin{aligned} \vec{D} &= \varepsilon_0 \vec{E} + \vec{p}_{\text{d}} = \varepsilon_0 \vec{E} + n\alpha_{\text{e}} \vec{E}_{\text{in}} = \varepsilon_0 \vec{E} + n\alpha_{\text{e}} \left( \vec{E} - \frac{\vec{p}_{\text{d,B}}}{3\varepsilon_0} \right) \\ &= \varepsilon_0 \left( 1 + \frac{n\alpha_{\text{e}}}{\varepsilon_0} \right) \vec{E} + \frac{n\alpha_{\text{e}}}{\varepsilon_0} \frac{\vec{D} - \varepsilon_0 \vec{E}}{3}. \end{aligned} \tag{4.62}$$

Here, we used the relation $\vec{p}_{\text{d}} = \vec{D} - \varepsilon_0 \vec{E} = -\vec{p}_{\text{d,B}}$. Taking into account (4.27), and collecting similar terms in (4.62), we obtain

$$\varepsilon = \frac{3 + \dfrac{2n\alpha_{\text{e}}}{\varepsilon_0}}{3 - \dfrac{n\alpha_{\text{e}}}{\varepsilon_0}}. \tag{4.63}$$

And the known Clausius–Mossotti relation follows:

$$\frac{n\alpha_{\text{e}}}{3\varepsilon_0} = \frac{\varepsilon - 1}{\varepsilon + 2}. \tag{4.64}$$

### 4.7.3 The dipole moment of polar molecules

As already mentioned, polar molecules have their own permanent dipole moments. In an external electric field, they acquire an additional induced dipole moment, determined by the electronic polarizability of molecules. In SI units, the permanent

**Table 4.2.** Permanent dipole moments of some polar molecules (in debye units: 1D = $3.336 \times 10^{-30}$ [C m]) [3].

| Molecule | $\mu_{d,0}$ | Molecule | $\mu_{d,0}$ |
|---|---|---|---|
| $H_2O$ | 1.85 | $NH_3$ | 1.47 |
| $SO_2$ | 1.62 | CO | 0.11 |
| NaCl | 8.5 | HCl | 1.08 |

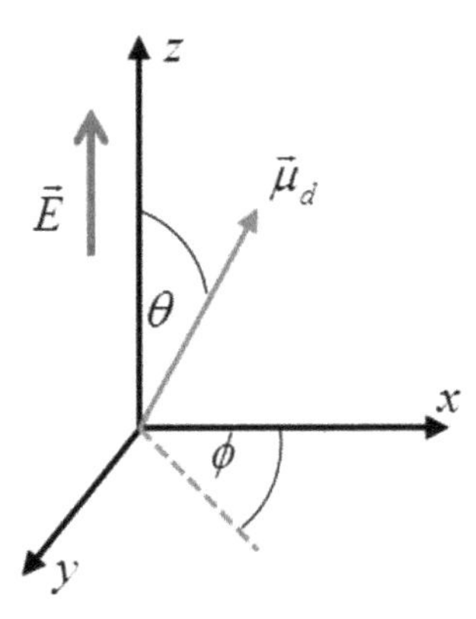


**Figure 4.14.** Polar coordinate system with the $z$-axis along the electric field vector $\vec{E}$.

dipole moment is measured in debye: 1 D = $3.33564 \times 10^{-30}$ [C m]. Table 4.2 shows the values for the permanent dipole moment $\mu_{d,0}$ of some typical polar molecules.

The polarization of polar dielectrics is associated with the external electric field's tendency to regulate the orientation of the electric moments of the molecules along itself. Dielectric polarization prevents random thermal motion, rotation, and mutual collisions of the molecules. Thus, polarization is determined by the ratio between the ordering influence of the electric field and the randomizing effect of thermal motion. Therefore, the polarization of polar dielectrics, in contrast to non-polar, should drop dramatically with increasing temperature.

Suppose that a dielectric contains $n_d$ molecules per unit volume with a permanent dipole moment whose amplitude is $\mu_{d,0}$. We introduce a polar coordinate system centered arbitrarily at a point of the dielectric with the polar $z$-axis directed along the external electric field vector $\vec{E}$ (figure 4.14). In this reference frame, the direction of the permanent component of the dipole moment of the arbitrary molecule is characterized by angle coordinates $\theta$ and $\phi$. In the absence of an external field, the directions of the dipole moments are isotropic.

The potential energy of a permanent dipole $\vec{\mu}_d$ in an electric field $\vec{E}$ is $U_\mu = -\vec{E} \cdot \vec{\mu}_d = -E\mu_{d,0}\cos(\theta)$. In order to find the distribution function of the directions of the molecules in the presence of an external field $\vec{E}$, we will use the

Boltzmann theorem, which states that in thermodynamic equilibrium the density distribution of polar molecules over $\theta$ and $\phi$ has the form

$$F_{\mathrm{E}}(\theta, \varphi) = \frac{n_{\mathrm{d}} \exp\left(\frac{E\mu_{\mathrm{d},0}}{k_{\mathrm{B}}T}\cos(\theta)\right)}{\int_0^{2\pi} \mathrm{d}\varphi \int_0^{\pi} \exp\left(\frac{E\mu_{\mathrm{d},0}}{k_{\mathrm{B}}T}\cos(\theta)\right)\sin(\theta)\mathrm{d}\theta}. \tag{4.65}$$

Here $k_{\mathrm{B}} = 1.38 \cdot 10^{-23}[\mathrm{J\ K^{-1}}]$ is the Boltzmann constant, $T$ is the temperature of the dielectric, $n_{\mathrm{d}}$ is the density of polar molecules.

We now calculate the dipole moment per unit volume of the dielectric. To do this, we multiply the distribution of (4.65) by $\mu_{\mathrm{d},0}\cos(\theta)$ and integrate with respect to the solid angle $\sin(\theta)\mathrm{d}\theta\mathrm{d}\varphi$:

$$p_{\mathrm{d,orient}} = \mu_{\mathrm{d},0} n_{\mathrm{d}} \frac{\int_0^{\pi} \exp\left(\frac{E\mu_{\mathrm{d},0}}{k_{\mathrm{B}}T}\cos(\theta)\right)\cos(\theta)\sin(\theta)\mathrm{d}\theta}{\int_0^{\pi} \exp\left(\frac{E\mu_{\mathrm{d},0}}{k_{\mathrm{B}}T}\cos(\theta)\right)\sin(\theta)\mathrm{d}\theta} = n_{\mathrm{d}}\mu_{\mathrm{d},0}\langle\cos(\theta)\rangle, \tag{4.66}$$

in which $\langle\cos(\theta)\rangle = \coth(\upsilon) - 1/\upsilon$ is the Langevin function and $\upsilon = \frac{E\mu_{\mathrm{d},0}}{k_{\mathrm{B}}T}$.

Figure 4.15 shows the dependence of the Langevin function on the parameter $\upsilon$. At $\upsilon \ll 1$, the density of the dipole moment is

$$p_{\mathrm{d,orient}} \approx \frac{n_{\mathrm{d}}\mu_{\mathrm{d},0}^2}{3k_{\mathrm{B}}T}E, \tag{4.67}$$

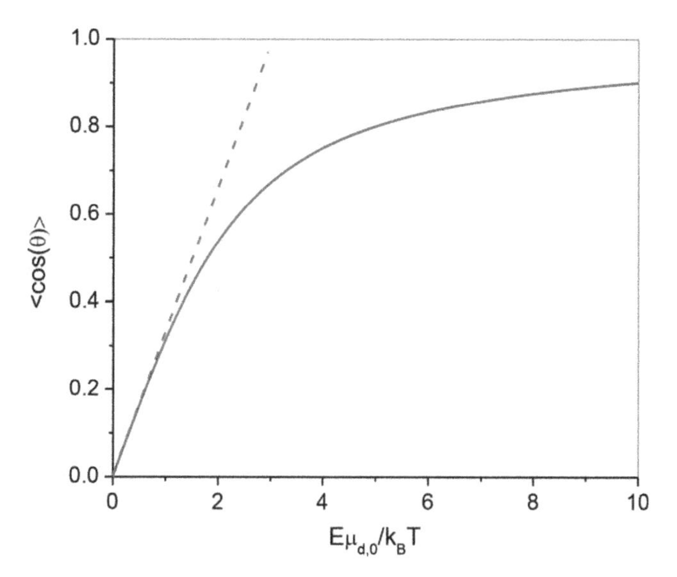


**Figure 4.15.** Dependence of the Langevin function $\langle\cos(\theta)\rangle$ on the parameter $\upsilon = E\mu_{\mathrm{d},0}/k_{\mathrm{B}}T$. The dashed line shows the low $\upsilon$ values as asymptotic.

so it increases linearly with the electric field. At $\upsilon \gg 1$, the dipole moment reaches its maximum of $\mu_{d,0}n$. In practice, only the $\upsilon \ll 1$ case is implemented, because breakdown and dissociation of molecules begin at lower values of the electric field in the dielectric, at which the linear dependence of the average dipole moment on the field still holds.

Introducing the averaged orientational polarizability

$$\alpha_{\text{orient}} = \frac{\mu_{d,0}^2}{3k_B T}, \tag{4.68}$$

we obtain the vector

$$\vec{p}_{d,\text{orient}} = n_d \alpha_{\text{orient}} \vec{E}. \tag{4.69}$$

Since polar molecules possesses electronic polarizability in addition to the orientation, it is convenient to introduce their effective polarizability:

$$\alpha_{\text{polar}} = \alpha_e + \alpha_{\text{orient}} = \alpha_e + \frac{\mu_{d,0}^2}{3k_B T}. \tag{4.70}$$

Note that, in general, one has to account for the presence of neighboring dipoles in liquid and solid polar dielectrics. Therefore, in the case of polar dielectrics, the Clausius–Mossotti formula (4.64) is valid only for gases and highly diluted solutions of polar liquids in non-polar solvents, for which it is transformed into the Langevin–Debye relation [4]:

$$\frac{n_d\left(\alpha_e + \dfrac{\mu_{d,0}^2}{3k_B T}\right)}{3\varepsilon_0} = \frac{\varepsilon - 1}{\varepsilon + 2}. \tag{4.71}$$

In [5] Onsager proposed a theory of liquid polar dielectric that takes into account the influence of neighboring dipoles. In it, he accounted for the fact that the field acting on the molecule from nearest neighbors depends on the direction of the dipole moment of the molecule. The molecule acquires an extra dipole moment induced by the field of neighboring molecules and this dipole moment, in turn, affects the neighboring molecules, changing their polarization state. In other words, the molecule itself is actively involved in the formation of the internal field acting on it [5]. We present the known Onsager formula for the permanent dipole moment of a molecule in the volume of a pure polar liquid without derivation:

$$\mu_{d,0}^2 = \frac{9k_B T}{4\pi n_d} \frac{(\varepsilon - \varepsilon_\infty)(2\varepsilon + \varepsilon_\infty)}{\varepsilon(\varepsilon_\infty + 2)^2}. \tag{4.72}$$

This equation allows one to compute the permanent dipole moment of a molecule $\mu_{d,0}$ using known values for the dielectric constant $\varepsilon$, density $n_d$, temperature $T$, and the upper limit of the high-frequency dielectric permittivity $\varepsilon_\infty \approx n^2$ ($n$ is the refractive index).

Onsager's theory agrees well with the experimental data of the permittivity of polar liquids in a wide range of temperatures, but liquids gives underestimates $\varepsilon$ for highly polar liquids. Kirkwood [6], Fröhlich [7], Cole [8], and many others have greatly developed and modified Onsager's theory. The modern theory of polarization of dielectrics can be found in [9–11], for example.

## 4.8 Ponderomotive forces in liquid dielectrics [12]

In electrodynamics, the forces acting on bodies and media from electric and magnetic fields are called ponderomotive forces. The term ponderomotive force was introduced a long time ago when, along with weighty bodies, people recognized the existence of weightless substances like ether and electrical fluid. In modern literature, it is generally accepted as the narrower definition of the forces acting on electric charges in oscillating inhomogeneous fields. Forces acting on dielectrics are considered just as electromagnetic forces acting on the free and bound charges. Following tradition, in this book we use the term ponderomotive forces for a more general class of phenomena, including the electrostrictive forces acting on dielectrics.

We will find the expression for the ponderomotive forces on a dielectric from (4.36) by analyzing variations in energy associated with an infinitesimal arbitrary (virtual) displacement $\vec{\xi}$, which continuously depends on the coordinates of the dielectric. Since the work of the ponderomotive force of density $\vec{f}_{\rm p}$ acting on volume $V$ equals $A = \int(\vec{f}_{\rm p} \cdot \vec{\xi})\mathrm{d}V = -\delta W$ and $\delta W = \frac{1}{2}\int\delta(\varepsilon_0\varepsilon E^2)\mathrm{d}V = \frac{1}{2}\int\delta(\frac{D^2}{\varepsilon_0\varepsilon})\mathrm{d}V$, then

$$\frac{1}{2}\int\delta\left(\vec{D}\cdot\vec{E}\right)\mathrm{d}V = \frac{1}{2\varepsilon_0}\int\delta\left(\frac{D^2}{\varepsilon}\right)\mathrm{d}V = -\int\left(\vec{f}_{\rm p}\cdot\vec{\xi}\right)\mathrm{d}V. \tag{4.73}$$

Expressing the value $\delta(D^2/\varepsilon)$ using the displacement $\vec{\xi}$, the vector that faces $\vec{\xi}$ on the left-hand side of (4.73) gives the ponderomotive force $\vec{f}_{\rm p}$ with the opposite sign, because the equality of (4.73) holds for all $\vec{\xi}$.

A variation of the electrostatic energy $\delta W$ is

$$\begin{aligned}\delta W &= \frac{1}{2\varepsilon_0}\int\delta\left(\frac{D^2}{\varepsilon}\right)\mathrm{d}V = \frac{1}{2\varepsilon_0}\int\left(2\frac{\vec{D}\cdot\delta\vec{D}}{\varepsilon} - \frac{D^2}{\varepsilon^2}\delta\varepsilon\right)\mathrm{d}V \\ &= \int\left(\vec{E}\cdot\delta\vec{D} - \frac{\varepsilon_0}{2}E^2\delta\varepsilon\right)\mathrm{d}V.\end{aligned} \tag{4.74}$$

The first integral on the right-hand side follows from the Gauss–Ostrogradsky theorem,

$$\begin{aligned}\int\left(\vec{E}\cdot\delta\vec{D}\right)\mathrm{d}V &= -\int\left(\vec{\nabla}\phi\cdot\delta\vec{D}\right)\mathrm{d}V \\ &= -\int\left(\vec{\nabla}\cdot\left(\phi\delta\vec{D}\right)\right)\mathrm{d}V + \int\phi\cdot\left(\vec{\nabla}\cdot\left(\delta\vec{D}\right)\right)\mathrm{d}V \\ &= -\int_S\phi\delta D_{\rm n}\mathrm{d}s + \int\phi\left(\vec{\nabla}\cdot\left(\delta\vec{D}\right)\right)\mathrm{d}V = \int\phi\delta\rho_{\rm f}\mathrm{d}V,\end{aligned} \tag{4.75}$$

in which we set the boundary surface location at infinity. Substituting (4.75) into (4.74), we obtain

$$\delta W = \int \phi \delta \rho_{\mathrm{f}} \mathrm{d}V - \frac{\varepsilon_0}{2} \int E^2 \delta \varepsilon \, \mathrm{d}V. \tag{4.76}$$

Now, we will find variations of the charge density $\delta\rho_{\mathrm{f}}$, the density of the dielectric $\delta\rho$, and the dielectric constant $\delta\varepsilon$. For simplicity, we assume that the virtual displacement occurs along the axis of an infinitely thin cylinder (figure 4.16).

The charge balance in volume *ABCD* is determined by the inside and outside charge flows through cross-sections *AB* and *CD*, which are, respectively, $\eta_{\mathrm{in}} = u(x)\rho_{\mathrm{f}}(x)$ and $\eta_{\mathrm{out}} = u(x + \mathrm{d}x)\rho_{\mathrm{f}}(x + \mathrm{d}x)$. A variation of the charge density in the considered volume $\mathrm{d}V = \mathrm{d}S\mathrm{d}x$ for a time interval $\delta t$ equals

$$\delta\rho_{\mathrm{f}} = \frac{(\eta_{\mathrm{in}} - \eta_{\mathrm{out}})\delta t \mathrm{d}S}{\mathrm{d}S\mathrm{d}x} = \frac{(\rho_{\mathrm{f}}(x)\xi(x) - \rho_{\mathrm{f}}(x + \mathrm{d}x)\xi(x + \mathrm{d}x))}{\mathrm{d}x} \approx -\frac{\partial(\rho_{\mathrm{f}}\xi)}{\partial x}, \tag{4.77}$$

where we took into account that $\xi(x) \approx v(x)\delta t$ and $\xi(x + \mathrm{d}x) \approx v(x + \mathrm{d}x)\delta t$ (figure 4.16).

During displacement, the dielectric constant in volume *ABCD* experiences a variation $\delta\varepsilon$ for two reasons. First, fluid flows into volume *ABCD* from other regions with different values of $\varepsilon$. Second, the density $\rho$ changes (figure 4.16) show that the displacement causes particles to move from volume $A'B'C'D'$ to volume *ABCD*. Although the total mass of the fluid is preserved in the process of displacement, the density changes since the interval $\mathrm{d}x'$ differs from $\mathrm{d}x$:

$$\rho' \mathrm{d}x' = (\rho + \delta\rho)\mathrm{d}x. \tag{4.78}$$

Hence,

$$\delta\rho = \frac{\mathrm{d}x' - \mathrm{d}x}{\mathrm{d}x}\rho,$$

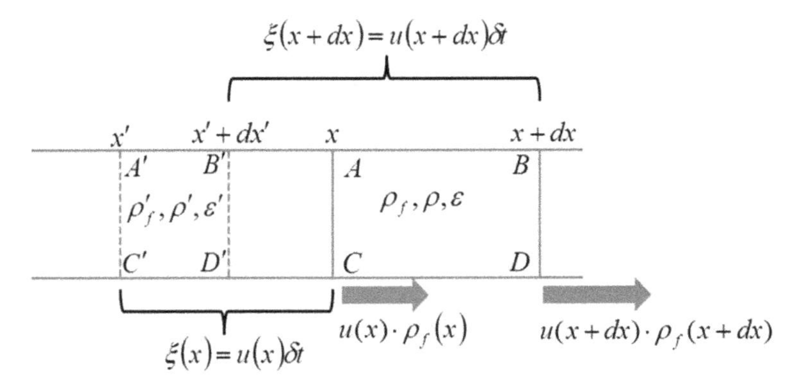


**Figure 4.16.** As a result of displacement $\xi$, the particles from volume $A'B'C'D'$ are displaced into volume *ABCD*.

or to second order accuracy

$$\delta\rho = -\frac{\partial\xi}{\partial x}\rho. \tag{4.79}$$

The fluid enters volume *ABCD* with a dielectric constant equal to

$$\varepsilon' + \frac{\partial\varepsilon}{\partial\rho}\delta\rho = \varepsilon' - \rho\frac{\partial\varepsilon}{\partial\rho}\frac{\partial\xi}{\partial x}. \tag{4.80}$$

Subtracting the unperturbed value of $\varepsilon$ from (4.47), we obtain

$$\delta\varepsilon = \varepsilon' - \varepsilon - \rho\frac{\partial\varepsilon}{\partial\rho}\frac{\partial\xi}{\partial x} = -\frac{\partial\varepsilon}{\partial x}\xi - \rho\frac{\partial\varepsilon}{\partial\rho}\frac{\partial\xi}{\partial x}. \tag{4.81}$$

Then, substituting (4.81) and (4.77) into (4.76), we obtain

$$\delta W = -\mathrm{d}S\int_{-\infty}^{+\infty}\phi\frac{\partial\left(\rho_f\xi\right)}{\partial x}\mathrm{d}x + \frac{\varepsilon_0}{2}\mathrm{d}S\int_{-\infty}^{+\infty}E^2\left(\frac{\partial\rho}{\partial x}\xi + \rho\frac{\partial\varepsilon}{\partial\rho}\frac{\partial\xi}{\partial x}\right)\mathrm{d}x. \tag{4.82}$$

Integrating by parts, and taking into account that all quantities become zero at the limits of integration, and that $E = E_x = -\partial\varphi/\partial x$, we obtain

$$\delta W = -\mathrm{d}S\int_{-\infty}^{+\infty}\left[\rho_\mathrm{f}E_x - \frac{\varepsilon_0}{2}\left(E^2\frac{\partial\varepsilon}{\partial x} - \frac{\partial}{\partial x}\left(E^2\frac{\partial\varepsilon}{\partial\rho}\rho\right)\right)\right]\xi\mathrm{d}x. \tag{4.83}$$

Since the work of ponderomotive forces at a virtual displacement along the $x$-axis is equal to

$$\delta A = \mathrm{d}S\int_{-\infty}^{+\infty}\xi f_{x,\mathrm{p}}\mathrm{d}x, \tag{4.84}$$

according to (4.73)

$$f_{x,\mathrm{p}} = \rho_\mathrm{f}E_x - \frac{\varepsilon_0}{2}\left(E^2\frac{\partial\varepsilon}{\partial x} - \frac{\partial}{\partial x}\left(E^2\rho\frac{\partial\varepsilon}{\partial\rho}\right)\right). \tag{4.85}$$

In general, the volumetric force acting on a compressible liquid dielectric in a non-uniform electric field (Helmholtz formula [12–15]) is

$$\vec{f}_\mathrm{p} = \rho_\mathrm{f}\vec{E} - \frac{\varepsilon_0}{2}\left(E^2\vec{\nabla}\varepsilon - \vec{\nabla}\left(E^2\rho\frac{\partial\varepsilon}{\partial\rho}\right)\right). \tag{4.86}$$

Thus, the equation of motion of an inviscid dielectric fluid in an electric field has the form

$$\rho\left(\frac{\partial\vec{u}}{\partial t} + \left(\vec{u}\cdot\vec{\nabla}\right)\vec{u}\right) = -\vec{\nabla}p + \rho_\mathrm{f}\vec{E} - \frac{\varepsilon_0}{2}\left(E^2\vec{\nabla}\varepsilon - \vec{\nabla}\left(E^2\rho\frac{\partial\varepsilon}{\partial\rho}\right)\right), \tag{4.87}$$

where $p$ is the hydrostatic pressure.

If there is no discontinuity in the dielectric fluid, then $\varepsilon$ is a function of the density $\rho$, and therefore $\vec{\nabla}\varepsilon = \frac{\partial\varepsilon}{\partial\rho}\vec{\nabla}\rho$. In this case,

$$\rho_0\left(\frac{\partial\vec{u}}{\partial t} + \left(\vec{u}\cdot\vec{\nabla}\right)\vec{u}\right) = -\vec{\nabla}p + \frac{\varepsilon_0}{2}\left(\vec{\nabla}\left(E^2\rho_0\frac{\partial\varepsilon}{\partial\rho}\right)\right). \tag{4.88}$$

Equation (4.88) assumes that the fluid is slightly compressible and contains no free charges.

For polar dielectrics [16, 17]

$$\rho_0\frac{\partial\varepsilon}{\partial\rho} = \alpha_E\varepsilon, \qquad \alpha_E = 1.3\text{–}1.6, \tag{4.89}$$

and for non-polar dielectrics [15]

$$\rho_0\frac{\partial\varepsilon}{\partial\rho} = \alpha_E\varepsilon, \qquad \alpha_E = \frac{(\varepsilon-1)\cdot(\varepsilon+2)}{3\varepsilon}, \tag{4.90}$$

and equation (4.88) takes the form

$$\rho_0\left(\frac{\partial\vec{u}}{\partial t} + \left(\vec{u}\cdot\vec{\nabla}\right)\vec{u}\right) = -\vec{\nabla}p + \frac{1}{2}\varepsilon_0\varepsilon\alpha_E\vec{\nabla}E^2 = -\vec{\nabla}\left(p - \frac{1}{2}\varepsilon_0\varepsilon\alpha_E E^2\right). \tag{4.91}$$

Thus, ponderomotive electrostrictive forces have a gradient nature, like hydrostatic pressure forces, but with the opposite sign. Therefore, liquid moves in a direction opposite to the gradient of the hydrostatic pressure, i.e. the hydrostatic pressure acts on the fluid as a compressing piston. But, the electric field stretches the fluid, and makes to move toward greater values of the square of the electric field.

In solid dielectrics, the only possible deformation is one that is not accompanied by macroscopic motion (electrostriction). Therefore, no matter how fast the electric field is switched on, if it is sufficiently large, cracks may occur in a solid dielectric. Everything is different in liquid dielectrics. If the electric field turns on slowly enough, the fluid moves under the influence of ponderomotive forces into the region of strong electric field, and the resulting difference in density compensates the action of the ponderomotive forces. If, on the other hand, it switches on fast enough in nanosecond and sub-nanosecond times, the fluid does not have time to move due to inertia. Therefore, at sufficiently large electric field gradients, the electrostrictive stresses arising in the fluid can cause ruptures of the fluid density (see chapters 1 and 5).

If the gradient of the electric field is not very large then, after some time, equilibrium is established at which the hydrostatic pressure gradient compensates the ponderomotive force:

$$\vec{\nabla}p_{\text{total}} = \vec{\nabla}(p + P_E) = 0, \qquad P_E = -\frac{1}{2}\alpha_E\varepsilon_0\varepsilon E^2 \tag{4.92}$$

and $p_{\text{total}} = p + P_E = p - \frac{1}{2}\alpha_E\varepsilon_0\varepsilon E^2 \approx p_0$, where $p_0$ is the hydrostatic pressure in the liquid, which is undisturbed by the electrostrictive (ponderomotive) forces.

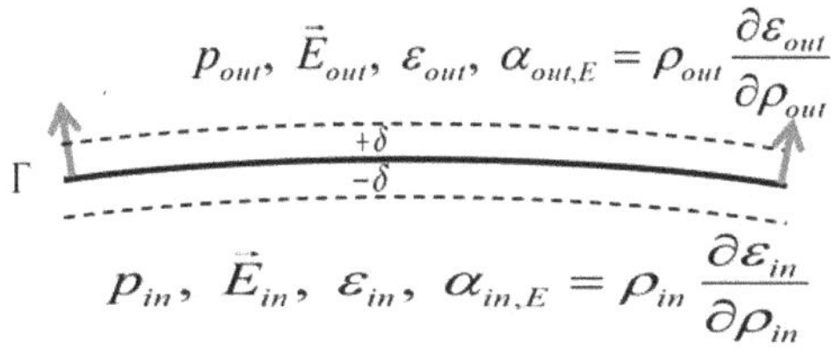

$p_{out}, \vec{E}_{out}, \varepsilon_{out}, \alpha_{out,E} = \rho_{out}\dfrac{\partial \varepsilon_{out}}{\partial \rho_{out}}$

$\Gamma$ $+\delta$ $-\delta$

$p_{in}, \vec{E}_{in}, \varepsilon_{in}, \alpha_{in,E} = \rho_{in}\dfrac{\partial \varepsilon_{in}}{\partial \rho_{in}}$

**Figure 4.17.** The boundary interface between two dielectrics. The arrows indicate the direction of the forces acting on the interface, as well as the direction along which the integration in formula (4.92) is performed.

## 4.9 Forces acting on the boundary between two dielectrics

As we note in chapter 1, applying a nanosecond voltage pulse to a sharp needle-like electrode in a dielectric fluid gives rise to volumetric ponderomotive forces that can create discontinuities in the liquid. In chapter 5, we provide the corresponding system of equations for calculations of fluid dynamics under the influence of ponderomotive forces.

The question arises of what will happen to the micropores: will they collapse under the action of surface tension or begins to expand? To answer this, we consider forces acting on the interface between two dielectrics. We denote the corresponding values of permittivity, density, hydrostatic pressure, and electric field for the inner and outer dielectrics as $\varepsilon_{\text{in}}$, $\rho_{\text{in}}$, $p_{\text{in}}$, and $\vec{E}_{\text{in}}$, and $\varepsilon_{\text{out}}$, $\rho_{\text{out}}$, $p_{\text{out}}$, and $\vec{E}_{\text{out}}$ (figure 4.17). The corresponding relations for the inner and outer dielectrics are $\rho_{\text{in}}\frac{\partial \varepsilon_{\text{in}}}{\partial \rho_{\text{in}}} = \alpha_{\text{in,E}}\varepsilon_{\text{in}}$ and $\rho_{\text{out}}\frac{\partial \varepsilon_{\text{out}}}{\partial \rho_{\text{out}}} = \alpha_{\text{out,E}}\varepsilon_{\text{out}}$, where the values $\alpha_{\text{in,E}}$ and $\alpha_{\text{out,E}}$ are given in (4.89) and (4.90) for polar and non-polar liquids. In the case of the vacuum pore (neglecting the vapor pressure), $p_{\text{in}} = 0$, $\varepsilon_{\text{in}} = 1$, and $\alpha_{\text{in,E}} = 0$.

We will find the force acting on a unit area of the interface without regard to surface tension forces, and assume that the surface charge $\rho_{\text{f}} = 0$. Integrating the right-hand side of equation (4.87) within small inner and outer regions along the normal to the interface (figure 4.17), we obtain

$$F_\Gamma = \int_{\Gamma-\delta}^{\Gamma+\delta}\left(-\frac{\partial p}{\partial l} - \frac{\varepsilon_0}{2}E^2\cdot\frac{\partial \varepsilon}{\partial l} + \frac{\varepsilon_0}{2}\frac{\partial}{\partial l}\left(\alpha_{\text{E}}\varepsilon E^2\right)\right)\mathrm{d}l. \tag{4.93}$$

By replacing $E^2$ in (4.92) with $E_{\text{t}}^2 + D_{\text{n}}^2/\varepsilon^2$, we obtain

$$F_\Gamma = \int_{\Gamma-\delta}^{\Gamma+\delta}\left(-\frac{\partial p}{\partial l} - \frac{\varepsilon_0}{2}E_{\text{t}}^2\frac{\partial \varepsilon}{\partial l} - \frac{\varepsilon_0}{2}D_{\text{n}}^2\frac{1}{\varepsilon^2}\frac{\partial \varepsilon}{\partial l} + \frac{\varepsilon_0}{2}\frac{\partial}{\partial l}\left(\alpha_{\text{E}}\varepsilon\left(E_{\text{t}}^2 + \frac{D_{\text{n}}^2}{\varepsilon^2}\right)\right)\right)\mathrm{d}l. \tag{4.94}$$

At the transition through the interface boundary, the tangential component of the electric field $E_{\text{t}}$ and the normal component of the induction $D_{\text{n}}$ are continuous ($E_{\text{in,t}} = E_{\text{out,t}}$, $D_{\text{in,n}} = D_{\text{out,n}}$). Then given that $\frac{1}{\varepsilon^2}\frac{\partial \varepsilon}{\partial l} = -\frac{\partial}{\partial l}\left(\frac{1}{\varepsilon}\right)$, (4.94) becomes

$$F_\Gamma = p_{\text{in}} - p_{\text{out}} + \frac{\varepsilon_0}{2}\left((\varepsilon_{\text{in}} - \varepsilon_{\text{out}})E_{\text{in,t}}^2 + \left(\frac{1}{\varepsilon_{\text{out}}} - \frac{1}{\varepsilon_{\text{in}}}\right)D_{\text{in,n}}^2 + (\varepsilon_{\text{out}}\alpha_{\text{out,E}} - \varepsilon_{\text{in}}\alpha_{\text{in,E}})E_{\text{in,t}}^2 + \left(\frac{\alpha_{\text{out,E}}}{\varepsilon_{\text{out}}} - \frac{\alpha_{\text{in,E}}}{\varepsilon_{\text{in}}}\right)D_{\text{in,n}}^2\right). \tag{4.95}$$

Substituting $D_{\text{in,n}} = \varepsilon_{\text{in}}E_{\text{in,n}}$ into (4.95) and grouping corresponding terms, we obtain

$$F_\Gamma = p_{\text{in}} - p_{\text{out}} + \frac{\varepsilon_0}{2}\left(\left(\varepsilon_{\text{in}}\left(1 - \alpha_{\text{in,E}}\right) - \varepsilon_{\text{out}}\left(1 - \alpha_{\text{out,E}}\right)\right)E_{\text{in,t}}^2 + \frac{\varepsilon_{\text{in}}}{\varepsilon_{\text{out}}}\left(\varepsilon_{\text{in}}\left(1 + \alpha_{\text{out,E}}\right) - \varepsilon_{\text{out}}\left(1 + \alpha_{\text{in,E}}\right)\right)E_{\text{in,n}}^2\right). \tag{4.96}$$

As we show in chapter 5, when an external electric field in the vicinity of the interface grows in a time of a few nanoseconds, the fluid does not have time to move, and therefore the contribution of $p_{\text{in}}$ and $p_{\text{out}}$ in $F_\Gamma$ can be neglected:

$$F_{\Gamma,\text{fast}} = \frac{\varepsilon_0}{2}\left(\left(\varepsilon_{\text{in}}\left(1 - \alpha_{\text{in,E}}\right) - \varepsilon_{\text{out}}\left(1 - \alpha_{\text{out,E}}\right)\right)E_{\text{in,t}}^2 + \frac{\varepsilon_{\text{in}}}{\varepsilon_{\text{out}}}\left(\varepsilon_{\text{in}}\left(1 + \alpha_{\text{out,E}}\right) - \varepsilon_{\text{out}}\left(1 + \alpha_{\text{in,E}}\right)\right)E_{\text{in,n}}^2\right). \tag{4.97}$$

On the other hand, when the field increases slowly, the induced fluid motion leads to growth of $p_{\text{in}}$ and $p_{\text{out}}$ to compensate the ponderomotive forces (chapter 5). In this case,

$$F_{\Gamma,\text{slow}} = \frac{\varepsilon_0}{2}(\varepsilon_{\text{in}} - \varepsilon_{\text{out}})\left(E_{\text{in,t}}^2 + \frac{\varepsilon_{\text{in}}}{\varepsilon_{\text{out}}}E_{\text{in,n}}^2\right). \tag{4.98}$$

## 4.10 Forces acting on a boundary of a dielectric sphere

In the case of a dielectric sphere, located in a weakly changing electric field in space ($R_0 \ll L_E$, where $R_0$ is the radius of the sphere and $L_E$ is the characteristic spatial scale of the electric field variation), the normal and tangential components of the field inside the sphere $E_{\text{in,t}}$ and $E_{\text{in,n}}$ are given by (4.46), and the corresponding force is

$$F_{\Gamma,\text{sphere}} = p_{\text{in}} - p_{\text{out}} + \frac{\varepsilon_0}{2}\left(\frac{3\varepsilon_{\text{out}}}{\varepsilon_{\text{in}} + 2\varepsilon_{\text{out}}}\right)^2 E_0^2 \cdot\left(\left(\varepsilon_{\text{in}}\left(1 - \alpha_{\text{in,E}}\right) - \varepsilon_{\text{out}}\left(1 - \alpha_{\text{out,E}}\right)\right)\sin^2(\theta) + \frac{\varepsilon_{\text{in}}}{\varepsilon_{\text{out}}}\left(\varepsilon_{\text{in}}\left(1 + \alpha_{\text{out,E}}\right) - \varepsilon_{\text{out}}\left(1 + \alpha_{\text{in,E}}\right)\right)\cos^2(\theta)\right), \tag{4.99}$$

where $E_0$ is the quasi-uniform electric field in the region of the outer dielectric.

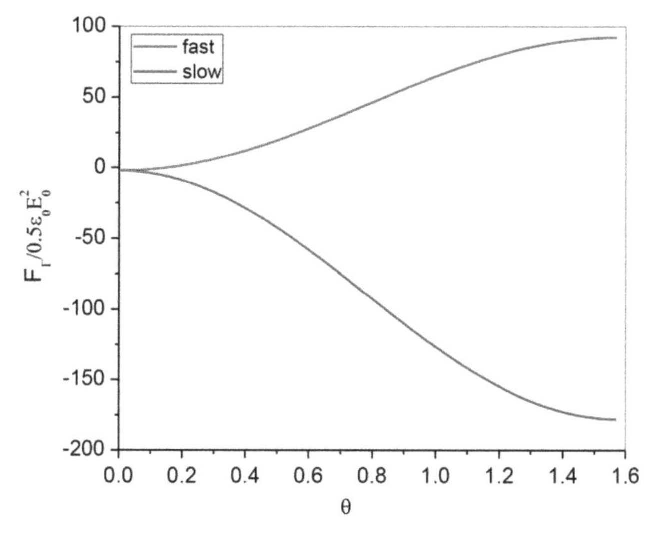


**Figure 4.18.** A vacuum pore in water. Forces acting on the boundary of the spherical vacuum pore in water (in units of $\varepsilon_0 E_0^2/2$) versus the azimuthal angle $\theta$. $\alpha_{\text{in,E}} = 0$, $\varepsilon_{\text{in}} = 1$, $\alpha_{\text{out},E} = 1.5$, and $\varepsilon_{\text{out}} = 81$.

Figure 4.18 shows the dependence of the forces acting on the boundary of a spherical pore in water, referred to as $\varepsilon_0 E_0^2/2$, versus the azimuthal angle $\theta$. It shows two limiting cases: when the field $E_0(t)$ changes quickly, the liquid does not have time to move, and therefore the hydrostatic pressure in (4.96) can be neglected, and the case of the field varies slowly and the hydrostatic pressure completely compensates the ponderomotive force.

Figure 4.18 shows that at slow growth of the electric field the force on the boundary is compressing ($F_\Gamma < 0$) at every azimuthal angle $\theta$, but with fast growth of the electric field the force is stretching ($F_\Gamma > 0$).

Similarly, figure 4.19 shows the forces acting on the boundary of a spherical vacuum pore in transformer oil versus $\theta$ in the two limiting cases and figure 4.20 shows the forces for a water droplet in transformer oil.

In chapter 6, we will discuss in detail the effect of surface tension and the induced flow of the forces acting on the bubble–water interface (bubble–oil), where we will consider the dynamics of vacuum pores in a pulsed non-uniform electric field.

We have seen that in the case of vacuum pores in water (figure 4.18), the absolute value of the force acting on the interface is minimum at the pole and maximum at the equator. Whereas for water droplets in oil, its maximum is at the pole and the minimum is at the equator (figure 4.20). In other words, the forces acting on a water droplet in oil tend to pull it along the electric field, regardless of how rapidly the electric field varies. It is known that, under the influence of a dc electric field, a drop of water in oil takes the shape of an ellipsoid elongated along the electric field (figure 4.21) (e.g. [19, 20]). Without going into detailed calculations, we estimate a force induced by the electric field on the interface boundary at a point $\theta = 0$.

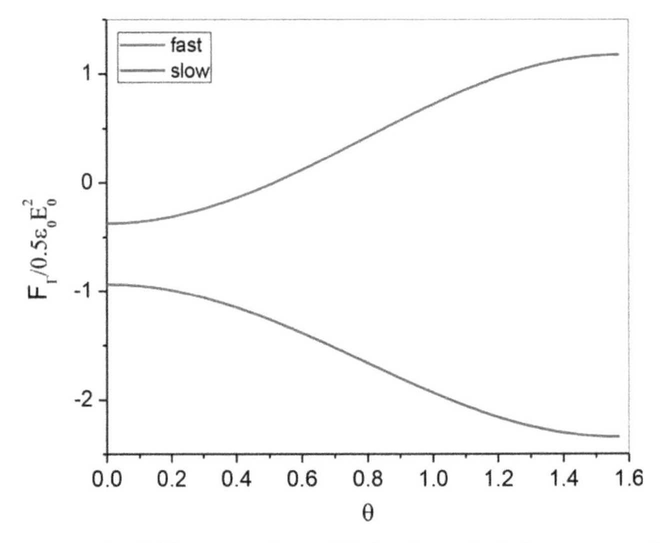


**Figure 4.19.** A vacuum pore in oil. The same as figure 4.18, but for a spherical vacuum pore in transformer oil, with $\varepsilon_{\text{in}} = 1$, $\alpha_{\text{in,E}} = 0$, $\varepsilon_{\text{out}} = 2.5$, and $\alpha_{\text{out,E}} = (\varepsilon_{\text{out}} - 1)(\varepsilon_{\text{out}} + 2)/3\varepsilon_{\text{out}}$.

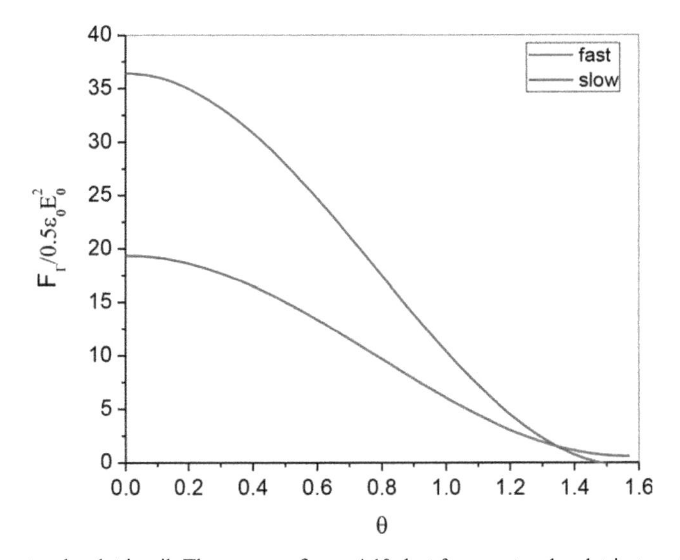


**Figure 4.20.** A water droplet in oil. The same as figure 4.18, but for a water droplet in transformer oil, with $\varepsilon_{\text{in}} = 81$, $\alpha_{\text{in,E}} = 1.5$, $\varepsilon_{\text{out}} = 2.5$, and $\alpha_{\text{out,E}} = (\varepsilon_{\text{out}} - 1)(\varepsilon_{\text{out}} + 2)/3\varepsilon_{\text{out}}$.

Following [13, 14], the electric field in a dielectric ellipsoid with semi-axes $a$ and $b$ (figure 4.21) is

$$E_{\text{in}} = \frac{\varepsilon_{\text{out}}}{\varepsilon_{\text{out}} + (\varepsilon_{\text{in}} - \varepsilon_{\text{out}})N_z} E_0, \tag{4.100}$$

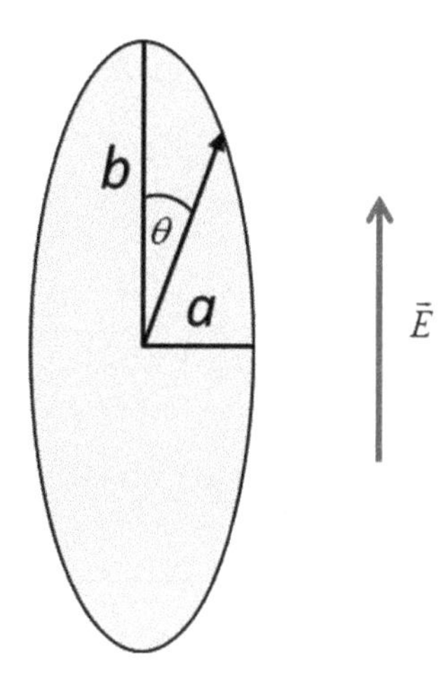


Figure 4.21. A water droplet in oil, deformed in an electric field.

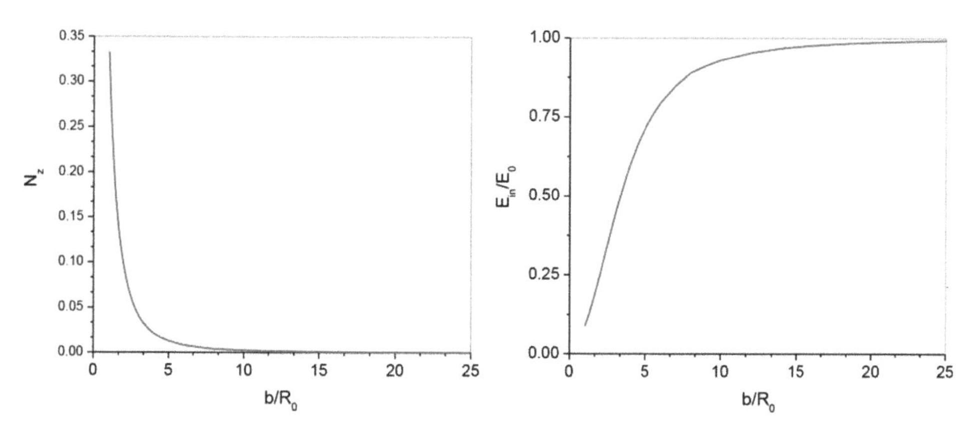


Figure 4.22. Dependence of $N_z$ and $E_{in}/E_0$ on $b/R_0$.

where

$$N_z = \frac{a^2 c}{2} \int_0^\infty \frac{ds}{\left(s + b^2\right)^{3/2} \left(s + a^2\right)}. \tag{4.101}$$

Due to the practical incompressibility, the volume of a stretching water droplet is maintained, and therefore the semi-axes $a$ and $b$ satisfy the condition $R_0^3 = a^2 b$, where $R_0$ is the initial radius of the spherical water droplet before exposure to the electric field. Expressing $a^2$ with $R_0$ and $b$, and substituting into (4.100), we obtain

$$N_z = \frac{1}{2} \int_0^\infty \frac{ds}{\left(s + (b/R_0)^2\right)^{3/2} (s + R_0/b)}. \tag{4.102}$$

At $b/R_0 = 1$ and $N_z = 1/3$, (4.100) coincides with (4.46). Figure 4.22 shows the dependences of $N_z$ and $E_{in}/E_0$ on $b/R_0$. As $b/R_0 \to \infty$, $N_z \to 0$ and $E_{in}/E_0 \to 1$.

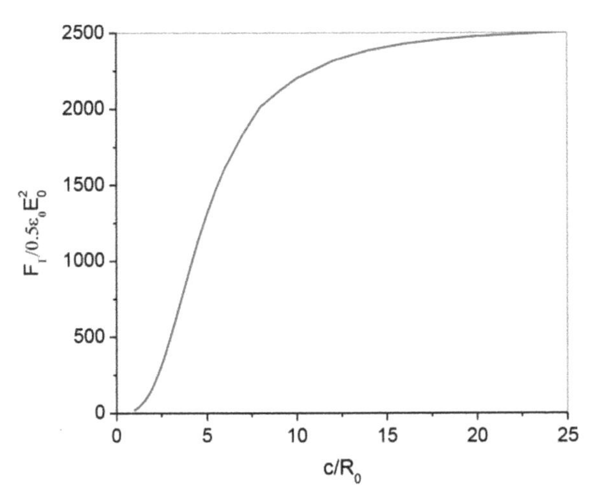


**Figure 4.23.** Dependence of $F_{\Gamma,\text{slow}}(\theta = 0)$ on $b/R_0$ for a drop of water in oil.

Since $E_{\text{in,t}} = 0$ at $\theta = 0$ and $E_{\text{in,n}}$ is given by (4.100), (4.98) becomes

$$F_{\Gamma,\text{slow}}(\theta = 0) = \frac{\varepsilon_0}{2}E_0^2\frac{\varepsilon_{\text{in}}}{\varepsilon_{\text{out}}}(\varepsilon_{\text{in}} - \varepsilon_{\text{out}})\left(\frac{\varepsilon_{\text{out}}}{\varepsilon_{\text{out}} + (\varepsilon_{\text{in}} - \varepsilon_{\text{out}})N_z}\right)^2. \tag{4.103}$$

Figure 4.23 shows the dependence of $F_{\Gamma,\text{slow}}(\theta = 0)$ on $b/R_0$.

At a fixed external field, $F_{\Gamma,\text{slow}}$ increases as $b/R_0$ increases, and then reaches an asymptotic saturated value at $b/R_0 > 20$:

$$F_{\Gamma,\text{slow, max}} = \frac{\varepsilon_0}{2}E_0^2\frac{\varepsilon_{\text{in}}}{\varepsilon_{\text{out}}}(\varepsilon_{\text{in}} - \varepsilon_{\text{out}}). \tag{4.104}$$

In fact, the stretching of the drop is limited by the surface tension force at the point $\theta = 0$:

$$F_{\text{s}} \approx -\frac{\sigma_{\text{s}}}{R_{\text{curv}}}, \tag{4.105}$$

where $R_{\text{curv}} = R_0\frac{R_0^2}{b^2}$ is the radius of curvature at $\theta = 0$. One can estimate the level of stretching of a water droplet in oil in a given static electric field by equating (4.105) to (4.104). In chapter 5, we will show that water droplets in oil under breakdown pulsed fields with fronts of several nanoseconds or less do not have enough time to stretch out noticeably and therefore their deviation from the spherical form can be neglected.

## References

[1] Pekker M and Shneider M N 2015 Pre-breakdown cavitation nanopores in the dielectric fluid in the inhomogeneous, pulsed electric fields *J. Phys. D: Appl. Phys.* **48** 424009

[2] CRC 2012 *CRC Handbook of Chemistry and Physics* 93th edn, ed W M Haynes (Boca Raton, FL: CRC Press)

[3] Israelachvili J N 1998 *Intermolecular and Surface forces* (New York: Academic)
[4] Debye P 1929 *Polar Molecules* (New York: Chemical Catalog Co)
[5] Onsager L 1936 Electric moments of molecules in liquids *J. Amer. Chem. Soc.* **58** 1486
[6] Kirkwood J G 1939 The dielectric polarization of polar liquids *J. Chem. Phys.* **7** 911
[7] Fröhlich H 1948 General theory of the static dielectric constant *Trans. Faraday Soc.* **44** 238
Fröhlich H 1958 *Theory of Dielectrics: Dielectric Constant and Dielectric Loss* (Oxford: Clarendon)
[8] Cole R H 1957 Induced polarization and dielectric constant of polar liquids *J. Chem. Phys.* **27** 33
[9] Hill N E 1969 *Dielectric Properties and Molecular Behaviour* (London: Van Nostrand Reinhold)
[10] Böttcher C J F 1973 *Theory of Electric Polarization* 2nd edn (Amsterdam: Elsevier)
[11] Scaife B K P 1998 *Principles of Dielectrics* 2nd edn (Oxford: Oxford University Press)
[12] Sivukhin D V 1996 *Electricity* (*The Course of General Physics* vol 3) (Moscow: Nauka, Fizmatlit) (in Russian)
[13] Landau L D and Lifshitz E M 1975 *Electrodynamics of Continuous Media* (*A Course of Theoretical Physics* vol 8) (Oxford: Pergamon)
[14] Tamm I E 1979 *Fundamentals of the Theory of Electricity* (Moscow: Mir)
[15] Panofsky W K H and Phillips M 1962 *Classical Electricity and Magnetism* (New York: Addison-Wesley)
[16] Jakobs J S and Lawson A W 1952 An analysis of the pressure dependence of the dielectric constant of polar liquids *J. Chem. Phys.* **20** 1161
[17] Ushakov V Y, Klimkin V F and Korobeynikov S M 2005 *Breakdown in Liquids at Impulse Voltage* (Tomsk: NTL) (in Russian)
Ushakov V Y, Klimkin F V and Korobeynikov S M 2005 *Impulse Breakdown of Liquids (Power Systems)* (Berlin: Springer)
[18] Miksis M J 1981 Shape of a drop in an electric field *Phys. Fluids* **24** 1967
[19] Sherwood J D 1988 Breakup of fluid droplets in electric and magnetic fields *J. Fluid. Mech.* **188** 133
[20] Zhang J, Zahn J D and Lin H 2013 Transient solution for droplet deformation under electric fields *Phys. Rev.* E **87** 043008

IOP Publishing

Liquid Dielectrics in an Inhomogeneous Pulsed Electric Field

M N Shneider and M Pekker

# Chapter 5

# Dynamics of a dielectric liquid in a non-uniform pulsed electric field

*System of equations and boundary conditions in prolate spheroidal coordinates. Numerical results and discussions. Flow arising at adiabatic switching of voltage and its rapid shutdown. Linearized equations and example results. Comparison of numerical results with measurements. Initiation of cavitation and nanosecond breakdown in oil on water microdroplets.*

## 5.1 System of equations and boundary conditions in prolate spheroidal coordinates

Consider the dynamics of a dielectric liquid (e.g. water) near a pointed needle-like electrode (figure 5.1) in a pulsed inhomogeneous electric field approximated by compressible fluid dynamics without discontinuities and the equation of state [1]. The standard set of mass and momentum conservation equations with the volumetric electrostrictive force (4.88) and the Tait equation of state (3.2) for compressible water have the form [2]

$$\frac{\partial \rho}{\partial t} + \nabla(\rho \vec{u}) = 0 \tag{5.1}$$

$$\rho\left(\frac{\partial \vec{u}}{\partial t} + (\vec{u} \cdot \nabla)\vec{u}\right) = -\nabla p + \frac{\varepsilon_0}{2}\nabla\left(E^2 \frac{\partial \varepsilon}{\partial \rho}\rho\right) + \rho\nu\left(\Delta \vec{u} + \frac{1}{3}\nabla(\nabla \cdot \vec{u})\right) \tag{5.2}$$

$$p = (p_0 + B)\left(\frac{\rho}{\rho_0}\right)^{\gamma} - B \tag{5.3}$$

with $\rho_0 = 1000\ \mathrm{kg\ m^{-3}}$, $p_0 = 10^5\ \mathrm{Pa}$, $B = 3.07 \cdot 10^8\ \mathrm{Pa}$, and $\gamma = 7.15$.

doi:10.1088/978-0-7503-1245-5ch5  

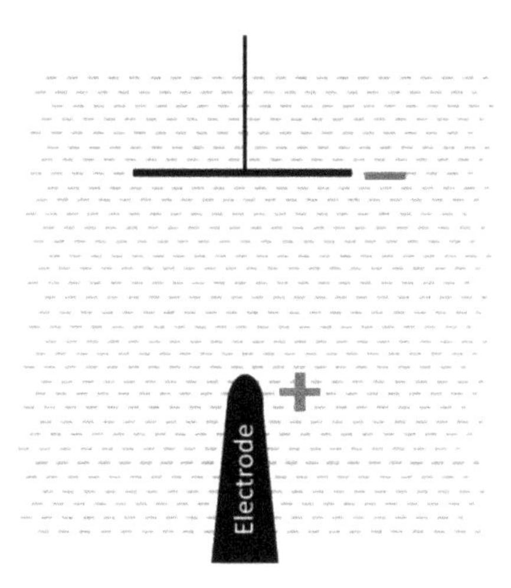


**Figure 5.1.** Scheme of the experiment. The electric field between the flat and needle-like electrodes applies electrostrictive tension on the fluid, which forces it to move into the region of maximum field (to the needle electrode, irrespective of the sign of the applied voltage).

Here $\rho$ is the fluid density, $p$ is the pressure, $\vec{u}$ is the velocity, $\nu$ is the kinematic viscosity, $\vec{E}$ is the electric field, $\varepsilon_0$ is the vacuum dielectric constant, and $\varepsilon$ is the liquid dielectric permittivity. For water at $T$ = 18 °C, $\varepsilon = 81$ (table 3.1), $\rho\ \partial\varepsilon/\partial\rho \approx 1.5\varepsilon$ (4.89), and $\nu \approx 10^{-6}$ m$^2$ s$^{-1}$ [3].

At the typical time scales for the voltage rise time of $\tau_0 \sim 1-10$ ns, the boundary layer $\Delta$ is much smaller than the radius of the electrode tip $r_{\rm el} \sim 5-100$ μm (the characteristic size of the strong electric field in the fluid). Indeed, since the boundary layer thickness is $\Delta \approx \sqrt{\nu t}$ [2], $\Delta \approx 0.1-0.3$ μm $\ll r_{\rm el}$. The characteristic time required to establish the boundary layer with a size comparable to the radius of curvature of the electrode $r_{\rm el}$ is 4–6 orders longer than the front rise time for the high-voltage pulse at the electrode: $\tau_0 \approx r_{\rm el}^2/\nu \sim 5\cdot 10^{-6}-10^{-2}$ s. Therefore, the terms related to viscosity can be neglected in equation (5.2). Due to the gradient form of the ponderomotive force, it is convenient to write (5.2) in the following form:

$$\rho\left(\frac{\partial\vec{u}}{\partial t} + (\vec{u}\cdot\nabla)\vec{u}\right) = -\nabla\left(p - \frac{\varepsilon_0}{2}E^2\frac{\partial\varepsilon}{\partial\rho}\rho\right) = -\nabla p_{\rm total}. \tag{5.4}$$

This means that the total pressure $p_{\rm total}$ acting on the dielectric liquid is a sum of the hydrostatic $p$ and the electrostriction-related $P_{\rm E} = -\alpha_{\rm E}\varepsilon_0\varepsilon E^2/2$ pressures. Reduction of the negative pressure in the fluid is associated with an increase in hydrostatic pressure (increasing of the density) due to the inflow of fluid into the region of negative pressure (retracting liquid in the region of maximum field).

We apply the standard set of boundary conditions for equations (5.1) and (5.4) in a cylindrical coordinate system with the axis along the electrode axis:

$$\begin{array}{lll} u_r|_\Gamma = 0 & u_z|_\Gamma = 0 & \left.\dfrac{\partial\rho}{\partial r}\right|_\Gamma = 0 \\ u_r|_{r=0} & \left.\dfrac{\partial u_z}{\partial r}\right|_{r=0} = 0 & \left.\dfrac{\partial\rho}{\partial r}\right|_{r=0} = 0 \\ \left.\dfrac{\partial u_r}{\partial r}\right|_{r=R} = 0 & u_z|_{r=R} = 0 & \left.\dfrac{\partial\rho}{\partial r}\right|_{r=R} = 0 \\ u_r|_{z=L} = 0 & \left.\dfrac{\partial u_z}{\partial z}\right|_{z=L} = 0 & \left.\dfrac{\partial\rho}{\partial z}\right|_{z=L} = 0 \end{array} . \tag{5.5}$$

Here $R$ and $L$ are the boundaries of the computational domain and $\Gamma$ is the electrode surface (figure 5.2). We assume the no-slip condition (the fluid velocity at the electrode goes to zero) and the continuity of the density and momentum fluxes on the boundaries of the computational domain.

Since we consider these processes in a highly inhomogeneous field in the vicinity of a sharp needle-like electrode that can be represented as a prolate ellipsoid, it is

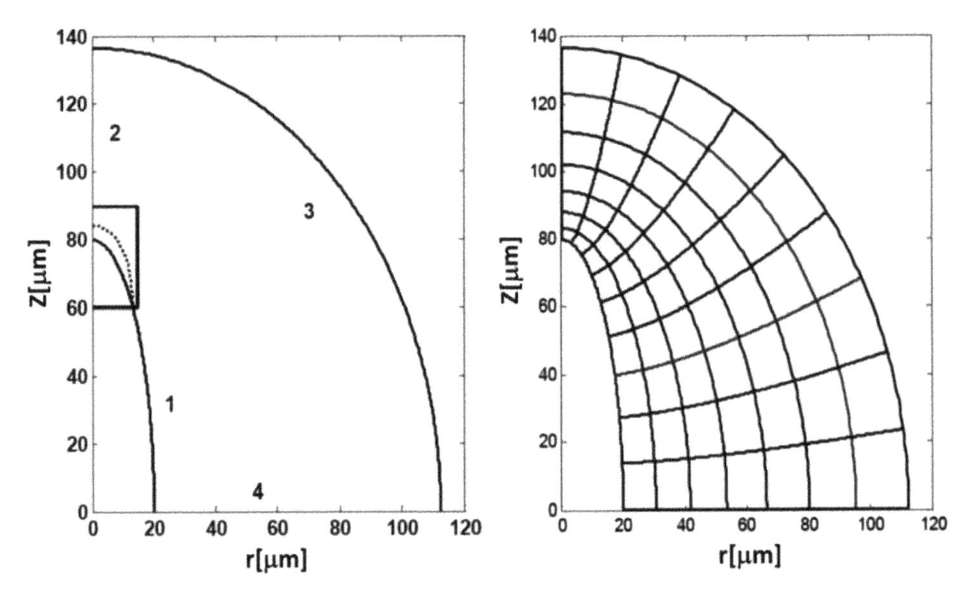


**Figure 5.2.** Boundaries of the area of integration of (5.1) and (5.4), and the grid in prolate spheroidal coordinates: 1 is the electrode surface, 2 is the axis of symmetry, and 3 and 4 are the boundaries of the computational domain. The area marked with a rectangle shows the two-dimensional distributions of pressure and velocity from figure 5.5. The dotted curve shows the region where the absolute value of the electrostrictive pressure exceeds the absolute value of the critical negative pressure necessary for cavitation. Reproduced with permission from [1]. Copyright 2013 the American Physical Society.

convenient to solve the equations for a compressible fluid in prolate spheroidal coordinates $\eta$ and $\mu$ [9]:

$$\begin{aligned} &r = a_e \cdot \sinh(\eta) \cdot \sin(\mu) \\ &z = a_e \cdot \cosh(\eta) \cdot \cos(\mu) \\ &\eta_0 \leqslant \eta < \infty \\ &0 \leqslant \mu \leqslant \pi/2 \end{aligned} . \tag{5.6}$$

The Jacobian of the transformation of variables $r$ and $z$ to $\eta$ and $\mu$ has the form

$$H = \left| \frac{\partial(r, z)}{\partial(\eta, \mu)} \right| = a_e^2 \left( \cosh(\eta) - \cos^2(\mu) \right). \tag{5.7}$$

With a sufficiently high accuracy, a needle-like electrode can be considered as an elongated ellipsoid with semiaxes $a_r$ and $a_z$. In this case, the eccentricity of the ellipsoid is $\xi_e = a_z/a_r > 1$ and, respectively, $a_e$ in (5.7) is

$$a_e = \sqrt{a_z^2 - a_r^2}. \tag{5.8}$$

On the boundary of the electrode,

$$\eta_0 = \text{arcth}(\xi_e). \tag{5.9}$$

The radius of curvature of the electrode's tip is

$$r_{el} = a_e \cdot \sinh^2(\eta_0)/\text{ch}(\eta_0). \tag{5.10}$$

Since the electrode is an equipotential, the Laplace equation for the potential in the poloidal coordinate system has the form

$$\Delta\Phi = \frac{1}{H^2} \frac{1}{\sinh(\eta)} \frac{\partial}{\partial\eta} \left( \sinh(\eta) \cdot \frac{\partial\Phi}{\partial\eta} \right) = 0. \tag{5.11}$$

Correspondingly,

$$\Phi = \Phi_0 \frac{\ln(\coth(\eta/2))}{\ln\left(\coth(\eta_0/2)\right)}, \tag{5.12}$$

in which $\Phi_0$ is the potential of the electrode. We assume the potential at infinity, i.e. the potential of the second electrode that is remote enough, to be zero. The expression for the electric field in the medium is given by

$$E_\eta = \frac{1}{\sqrt{H}} \frac{\partial\Phi}{\partial\eta} = \frac{1}{\sqrt{H} \cdot \sinh(\eta)} \cdot \frac{\Phi_0}{\ln\left(\coth(\eta_0/2)\right)}. \tag{5.13}$$

Then, the system of equations (5.1) and (5.2) in the poloidal coordinate system reduces to:

$$\begin{aligned}
&\frac{\partial\rho}{\partial t}+\frac{1}{H^2}\left(\frac{1}{\sinh(\eta)}\frac{\partial}{\partial\eta}(H\cdot\sinh(\eta)\cdot\rho u_\eta)\right.\\
&\qquad\left.+\frac{1}{\sin(\mu)}\frac{\partial}{\partial\mu}(H\cdot\sin(\mu)\cdot\rho u_\mu)\right)=0\\
&\frac{\partial u_\eta}{\partial t}-\frac{1}{H^2}\left(\frac{\partial}{\partial\eta}(H\cdot u_\mu)-\frac{\partial}{\partial\mu}(H\cdot u_\eta)\right)\cdot u_\mu+\frac{1}{H}\frac{\partial}{\partial\eta}\left(\frac{u_\mu^2+u_\eta^2}{2}\right)\\
&\qquad+\frac{1}{\rho}\frac{1}{H}\frac{\partial}{\partial\eta}(p-\alpha\varepsilon\varepsilon_0E^2)=0\\
&\frac{\partial u_\mu}{\partial t}+\frac{1}{H^2}\left(\frac{\partial}{\partial\eta}(H\cdot u_\mu)-\frac{\partial}{\partial\mu}(H\cdot u_\eta)\right)\cdot u_\eta+\frac{1}{H}\frac{\partial}{\partial\mu}\left(\frac{u_\mu^2+u_\eta^2}{2}\right)\\
&\qquad+\frac{1}{\rho}\frac{1}{H}\frac{\partial}{\partial\mu}(p-\alpha\varepsilon\varepsilon_0E^2)=0
\end{aligned}\tag{5.14}$$

In accordance with (5.5), the set of boundary conditions for this system in the poloidal coordinate system is

$$\begin{array}{llll}
u_\mu=0,\ u_\eta=0,\ \partial\rho/\partial\eta & \text{at } \eta=\eta_0 & 0\leqslant\mu\leqslant\pi/2 \\
u_\mu=0,\ \partial u_\eta/\partial\eta=0,\ \partial\rho/\partial\eta=0 & \text{at } \eta=\eta_1 & 0\leqslant\mu\leqslant\pi/2 \\
u_\mu=0,\ \partial u_\eta/\partial\mu=0,\ \partial\rho/\partial\mu=0 & \text{at } \eta_0\leqslant\eta\leqslant\eta_1 & \mu=0 \\
\partial u_\eta/\partial u_\eta=0,\ \partial u_\eta/\partial\mu=0,\ \partial\rho/\partial\mu=0 & \text{at } \eta_0\leqslant\eta\leqslant\eta_1 & \mu=\pi/2.
\end{array}\tag{5.15}$$

The time dependent system of equation (5.14), together with the equation of state (5.3) and the boundary conditions (5.15), was solved in [1] using a MacCormack second-order scheme [5].

## 5.2 Numerical results and discussion [1]

We consider the transition process in distilled water when a voltage is applied to the needle-plane electrode system. We choose the following set of parameters in our calculations: a dielectric permittivity of $\varepsilon = 81$, $\alpha_E = 1.5$, a negative pressure threshold at which cavitation starts of −30 MPa, a radius of the electrode tip of $r_{el} = 5$ μm, and an eccentricity (ratio of the major semiaxis of the prolate ellipsoid to its small semiaxis) of $\xi = 5$. Figure 5.2 presents the geometry of the problem and the grid in the prolate spheroidal coordinates. In all computed cases, the linear form of the voltage pulse is $U(t) = U_0t/t_0$ with $t \leqslant \tau_0$, in which $U_0 = 7$ kV is the maximal voltage on the electrode and $\tau_0$ is the front duration. To study the effect of voltage rise time, we perform calculations for $\tau_0 = 1$, 5, 10, and 15 ns. Figure 5.3 shows the negative pressure in the fluid along the symmetry axis ($r = 0$, $z$) caused by the electrostrictive forces at different moments of the voltage pulse.

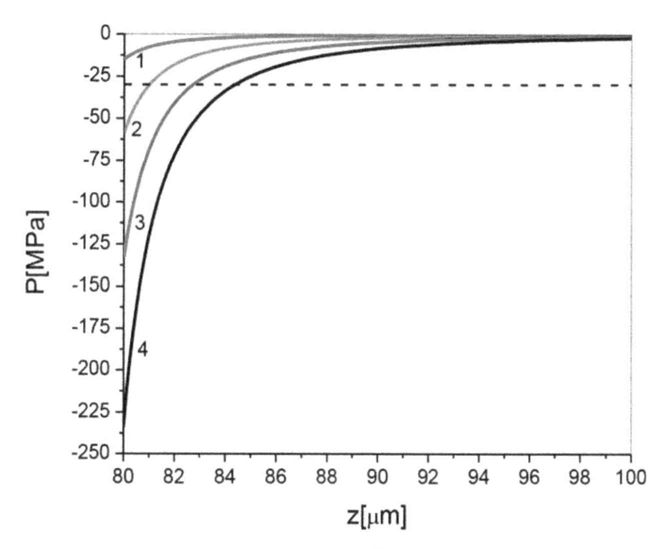


**Figure 5.3.** Electrostrictive negative pressure $P_E = -\alpha_E \varepsilon \varepsilon_0 E^2/2$ in the fluid along the axis of symmetry ($r = 0, z$) at the time moments $t/\tau_0 = 0.25$ (curve 1), 0.5 (curve 2), 0.75 (curve 3), and 1 (curve 4). The dashed line shows the pressure threshold for cavitation when a rupture in the continuity of the fluid occurs. Reproduced with permission from [1]. Copyright 2013 the American Physical Society.

Electrostrictive forces cause the fluid to flow to the electrode. As a result, the absolute value of the total pressure in the fluid is $|p_{tot}| = |p + P_E| < |P_E|$, $P_E = -\alpha_E \varepsilon \varepsilon_0 E^2/2$. Thus, at relatively sharp voltage pulse fronts conditions arise for the development of cavitation. Under these conditions, cavitation cannot arise at more gentle pulse fronts. This is clearly demonstrated by the computed total pressure distributions on the symmetry axis at different times as shown in figure 5.4(a)–(d), in which the dashed line shows the threshold pressure for cavitation.

In [6] it was shown that the size of the area of negative pressure, where the conditions for fluid cavitation ruptures are satisfied, is proportional to the square of the applied voltage amplitude, and decreases inversely proportionally to the fourth power of the radius of the needle-like electrode's tip. The calculations we present here are in agreement with these qualitative regularities.

The velocity of the fluid flow arising under the considered conditions during the entire voltage pulse remains subsonic and does not exceed tens of meters per second (the middle column of figure 5.4). Fluid influx to the electrode causes changes in density. However, in all computed cases, the maximum change in fluid density in the vicinity of the electrode does not exceed a few percent (the right column of figure 5.4).

The obtained results show a qualitative difference between the behaviors of the liquid at relatively fast and slow rises of a non-uniform electric field. At short rise times, there are large tensile stresses (large negative pressure), which can lead to discontinuities and cavitation formation of nanovoids. At relatively slow increases of the field, the rising flow leads to a strong decrease of the negative pressure down to values below the cavitation threshold, so fluid ruptures do not occur. At $\tau_0 \approx 1$ ns,

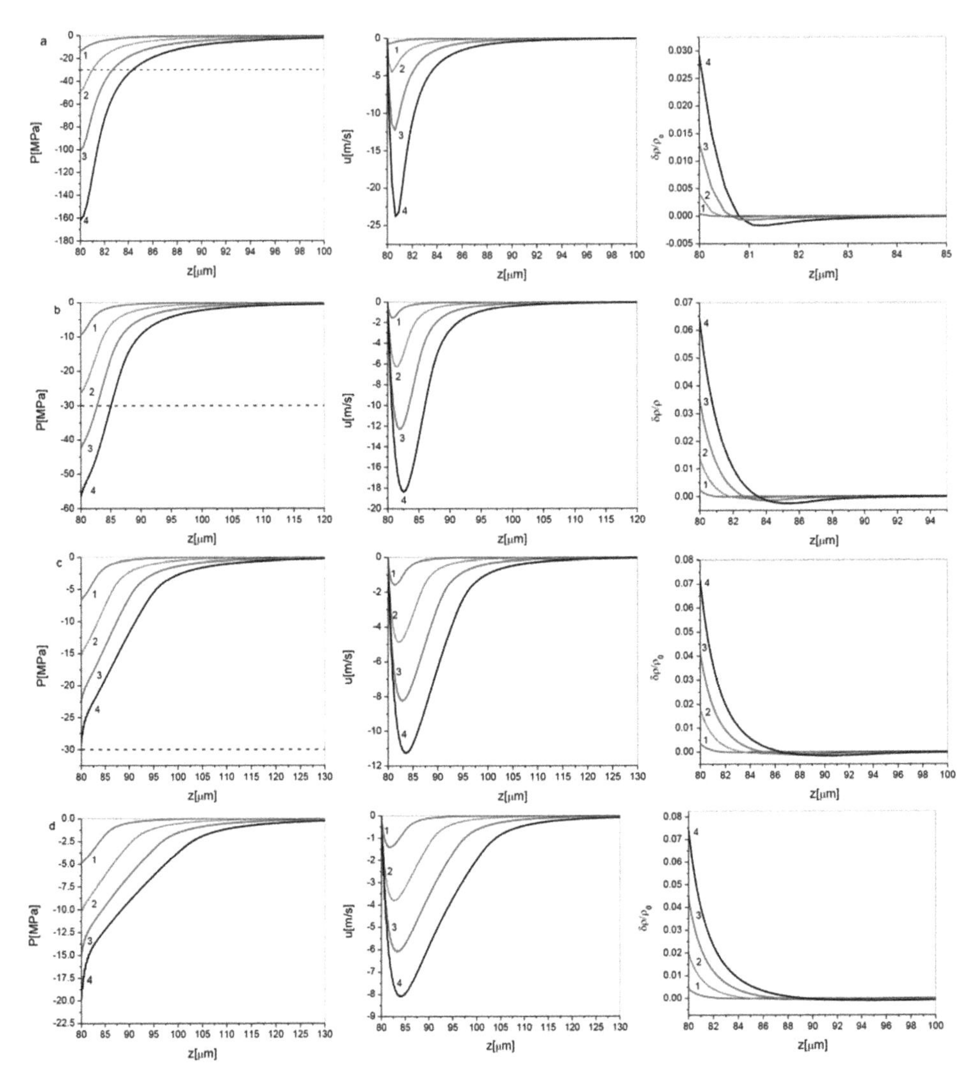

**Figure 5.4.** Longitudinal distributions of the total pressure $p_{\rm tot}$, flow velocity $u_z$, and the relative density perturbation $\delta\rho/\rho_0$ along the symmetry axis ($r = 0$, $z$) for a pulse with rise time $\tau_0 = 1$ ns (a), $\tau_0 = 5$ ns (b), $\tau_0 = 10$ ns (c), and $\tau_0 = 15$ ns (d). Curve 1 corresponds to the time moment $t/\tau_0 = 0.25$, 2 to $t/\tau_0 = 0.5$, 3 to $t/\tau_0 = 0.75$, and 4 to $t/\tau_0 = 1$. The dashed line shows the pressure threshold for cavitation when a rupture in the continuity of the fluid occurs. Reproduced with permission from [1]. Copyright 2013 the American Physical Society.

the absolute value of the negative pressure associated with the ponderomotive forces is much greater than the hydrostatic pressure. However, at $\tau_0 > 5$ ns, the hydrodynamic pressure almost completely compensates for the electrostrictive pressure. Note that this result is consistent with the earlier estimate (1.40) $\tau_0 \approx \tau_{\rm P} \approx 1.55\frac{r_{\rm el}}{c_{\rm S}} \approx 10^{-3} r_{\rm el} \approx 5 \cdot 10^{-9}$ s, in which $r_{\rm el}$ is given in meters.

Figure 5.5 shows contours of the total pressure $p_{\rm tot}$, the ponderomotive pressure $p_{\rm E}$, and the relative density $\delta = (\rho - \rho_0)/\rho_0$ at $t = 5$ ns for a pulse with a front

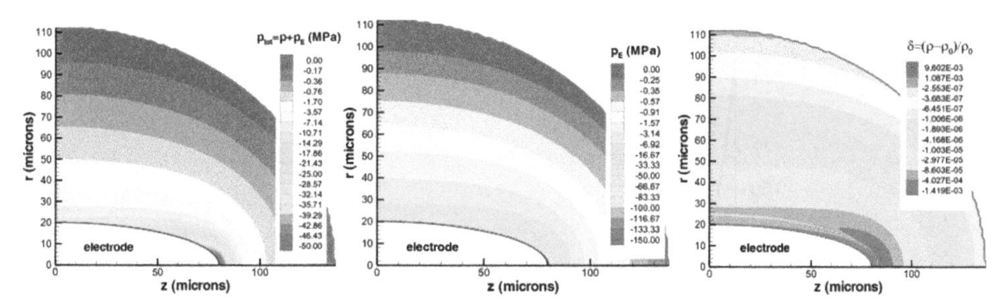

**Figure 5.5.** Contours of the total pressure (left), ponderomotive pressures (middle), and the relative density perturbation (right) at $t = 5$ ns for a pulse with a front duration of $\tau_0 = 5$ ns. Reproduced with permission from [1]. Copyright 2013 the American Physical Society.

duration of $\tau_0 \approx 5$ ns. The region of the electrode with absolute values of negative pressure greater than 30 Mpa, where fluid rupture (formation of cavitation nanovoids) can form (left plot) is relatively small and extends from the vicinity of the electrode tip to a distance of about 5 μm. The fluid density perturbation (right plot) is maximal in the vicinity of the electrode surface, and just outside it a small region of rarefaction occurs due to fluid motion and stretching under the influence of ponderomotive forces.

At the same electrode geometry and voltage parameters, pulse cavitation ruptures cannot occur in a dielectric fluid with a dielectric constant much lower than that of water. This is because the tensile stresses, determined by the electrostriction force in (5.2), are linearly dependent on the dielectric constant of the liquid.

## 5.3 Flow arising at adiabatic switching of voltage and its rapid shutdown [1]

If the voltage on the electrode is switching slowly enough, the flow occurring in the fluid has time to reduce the total pressure to such an extent that cavitation ruptures cannot occur. In this case, the hydrostatic pressure at the electrode can reach the value $p = |P_E|$. At a sharp turn-off of the applied voltage the electrostriction pressure disappears and a large gradient of the hydrostatic pressure leads to the formation of fluid flow away from the electrode. As a result, due to inertia of the fluid flow, the formation of negative pressure regions and cavitation ruptures is possible.

Figure 5.6 shows the formation of negative pressure near the electrode after a sharp (instantaneous) shutdown of the electric field. As the initial condition, the hydrostatic pressure is assumed to be equal to the absolute value of the electrostrictive pressure $|P_E| = \frac{1}{2}\alpha_E\varepsilon\varepsilon_0 E^2$ at the maximum voltage on the electrode $U_0$. Figure 5.6 clearly shows that a region of negative pressure forms in the vicinity of the electrode within a time of about a few nanoseconds.

## 5.4 Linearized equations and example results [1]

As the calculations show, changes in the fluid density do not exceed a few per cent, and the resulting flow rate is much less than the velocity of sound; so the system of

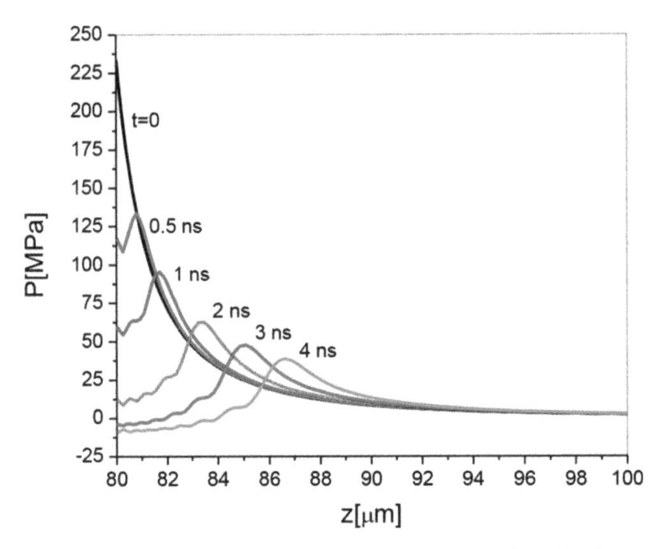


**Figure 5.6.** Longitudinal distributions of the hydrostatic pressure along the axis of symmetry ($r = 0$, $z$) at different time moments after a voltage interruption. Reproduced with permission from [1]. Copyright 2013 the American Physical Society.

equations (5.1), (5.3), (5.4) can be simplified by linearization. In the spherically symmetric case, it is reduced to the form [1]

$$\begin{cases} \dfrac{\partial u}{\partial t} = -\dfrac{1}{\rho_0}\dfrac{\partial}{\partial r}\left(p - \dfrac{\alpha_E}{2}\varepsilon_0\varepsilon E^2\right) \\ \dfrac{\partial p}{\partial t} = -\rho_0 c_s^2 \dfrac{1}{r^2}\dfrac{\partial}{\partial r}(r^2 u) \\ c_s = \sqrt{\dfrac{B\gamma}{\rho_0}} \approx 1500\ \mathrm{ms}^{-1} \\ E = U(t)\dfrac{r_{el}}{r^2} \end{cases} . \tag{5.16}$$

Figure 5.7 shows the total pressure and fluid velocity for a spherical electrode with a radius of 5 μm and voltage pulses with rise times of 1 ns, 5 ns, 10 ns, and 15 ns obtained by solving this simplified system of equations. We set the voltage amplitude on the electrode to $U_0 \approx 3.32$ kV, so that the electric field on a spherical electrode was equal to the field at the end of an ellipsoidal electrode. These results show good agreement with the two-dimensional calculation results shown in figure 5.4 for pulses with short fronts (1 ns and 5 ns), and a reasonable agreement for longer pulses. This is related to the fact that the sizes of the negative pressure region are different in the cases of an ellipsoidal and a spherical electrode.

## 5.5 Comparison of numerical results with measurements [7]

We will not go into the details of the experiment in question, because they are described in many papers (e.g. [8–10]) and will focus only on the idea of the method

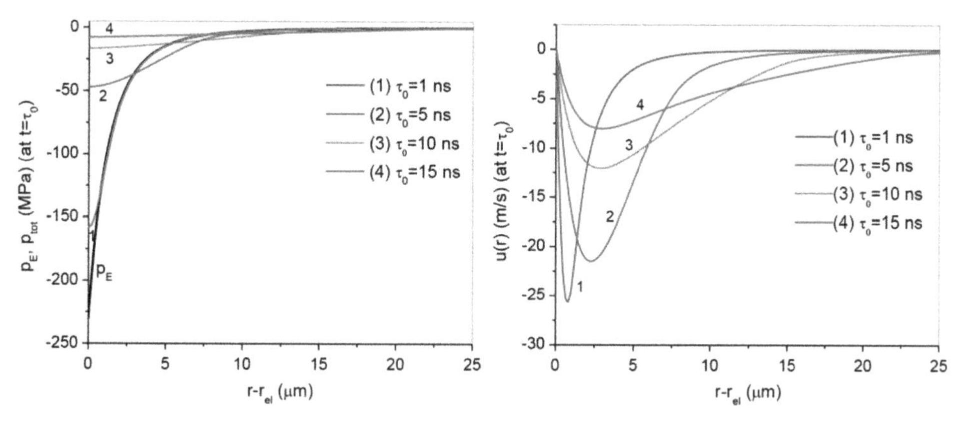


**Figure 5.7.** Radial distributions of the electrostriction and total pressures (left) and the fluid velocity (right) for a spherical electrode of radius 5 μm at the time $t = \tau_0$ for voltage pulses with fronts $\tau_0 = 1$ ns, 5 ns, 10 ns, and 15 ns obtained by calculation of the linearized system (5.16). Reproduced with permission from [1]. Copyright 2013 the American Physical Society.

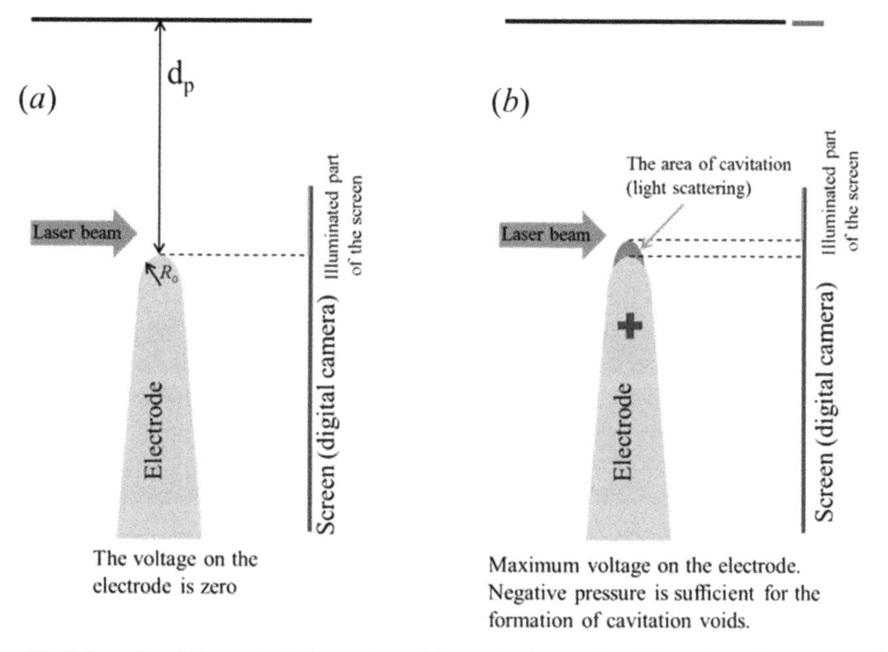


**Figure 5.8.** Schematic of the optical observation of the cavitation region. The radius of curvature of the electrode is $r_e = 35$ μm. The distance between the needle tip and the flat electrode is $d_p = 1.5$ mm. At (a) the voltage between the electrodes is zero and at (b) the voltage is at a maximum. Reproduced from [7].

of detection of void formation, proposed in [11]. Since the region containing voids leads to Raleigh scattering of light (see chapter 8), we can easily define the cavitation boundary by subtracting the illumination on the screen with voltage applied (figure 5.8(b)) from the illumination on the screen with no voltage (figure 5.8(a)).

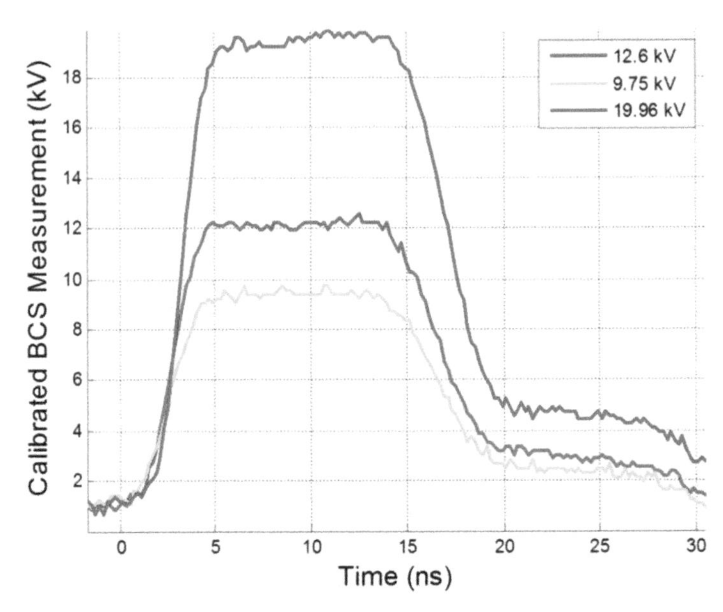


**Figure 5.9.** Time dependence of voltage on the electrode. Reproduced from [7].



**Figure 5.10.** Schlieren images of the experiment in [7] for the noted voltage pulse strengths and corresponding negative pressures. Each image is a 1 ns exposure starting at the noted times. Reproduced from [7].

The Schlieren method, which is a modification of the shadow [12], was used in [13–15]. Figure 5.9 presents the profiles of the voltage supplied to the electrode [7]. Figures 5.10 and 5.11 show Schlieren images (of size 340 × 230 μm). The 'dark' area in the vicinity of the electrode shows cavitation. There is less cavitation for voltage $U_0 = 12.6$ kV than at $U_0 = 19.96$ kV; and at $U_0 = 0.72$ kV cavitation is totally absent. In all of our measurements, we did not observe any emission in the vicinity of the electrode. This means that in all of our measurements we were always below the threshold of breakdown.

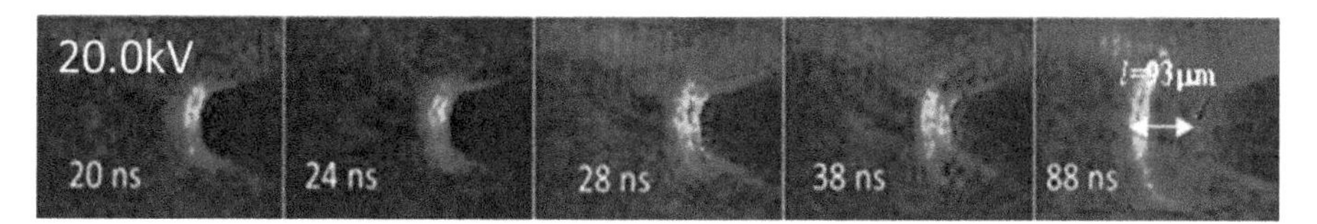


**Figure 5.11.** Schlieren images for 20 kV at later 1 ns exposures starting at 20 ns, 24 ns, 28 ns, 38 ns, and 88 ns. Reproduced from [7].

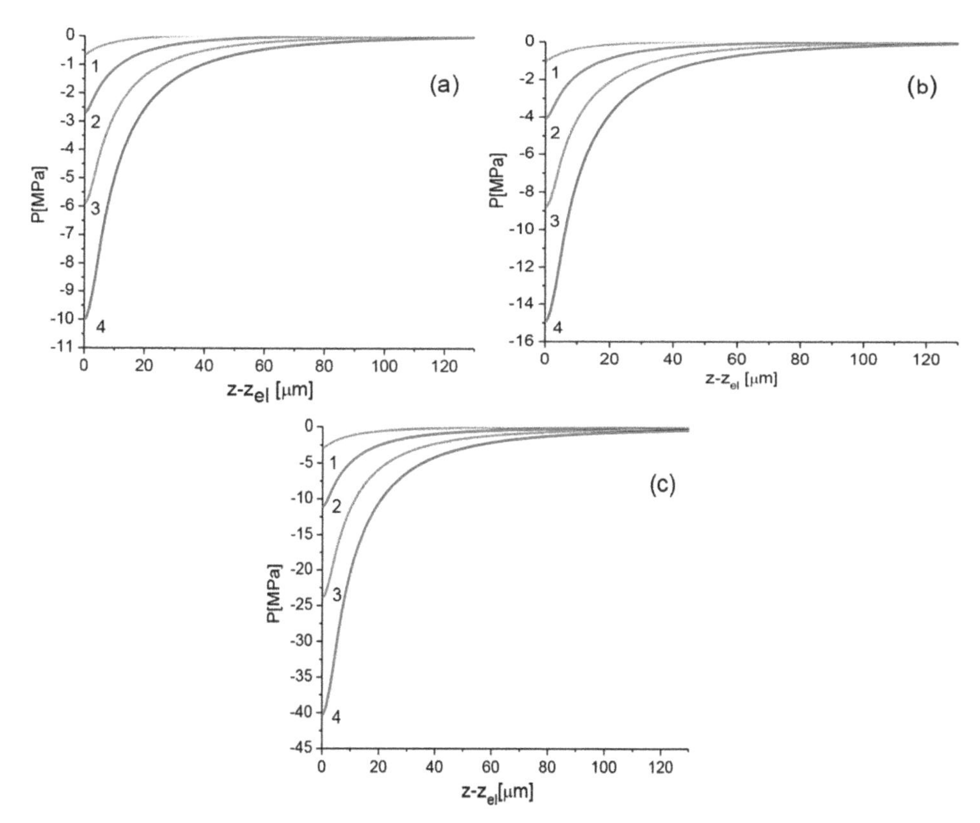


**Figure 5.12.** Distributions of total pressure along the symmetry axis of the electrode. Plot (a) corresponds to a $U_0$ of 10 kV, (b) to 12.2 kV, and (c) to 20 kV. Curve 1 shows the results at $t = 1$ ns, curve 2 at 2 ns, curve 3 at 3 ns, and curve 4 at 4 ns. $z_{\text{el}}$ is the $z$ coordinate value of the edge of the sharp electrode. Reproduced from [7].

Figure 5.12 presents the results of calculating equation (5.14) with boundary conditions (5.15) and equation of state (5.3) for the same parameters of the experiment described in section 5.5. The radius of curvature of the needle electrode is $R_0 = 35$ μm, and the amplitude of the voltage on the electrode is $U_0 = 20$ kV, 12.2 kV, and 10.0 kV. The time dependence of the voltage is varied in accordance with figure 5.9: during time interval 0–4 ns, the voltage grows linearly, during 4–14 ns it is constant, during 14–18 ns it drops linearly, and during 18–150 ns it is zero.

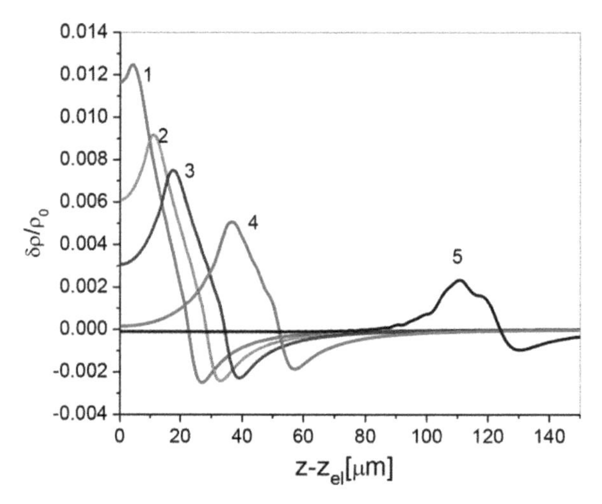


**Figure 5.13.** Distributions of perturbation density $\delta\rho/\rho_0$ along the symmetry axis of the electrode. Curve 1 corresponds to $t = 20$ ns, curve 2 to 24 ns, curve 3 to 28 ns, curve 4 to 40 ns, and curve 5 to 88 ns. Reproduced from [7].

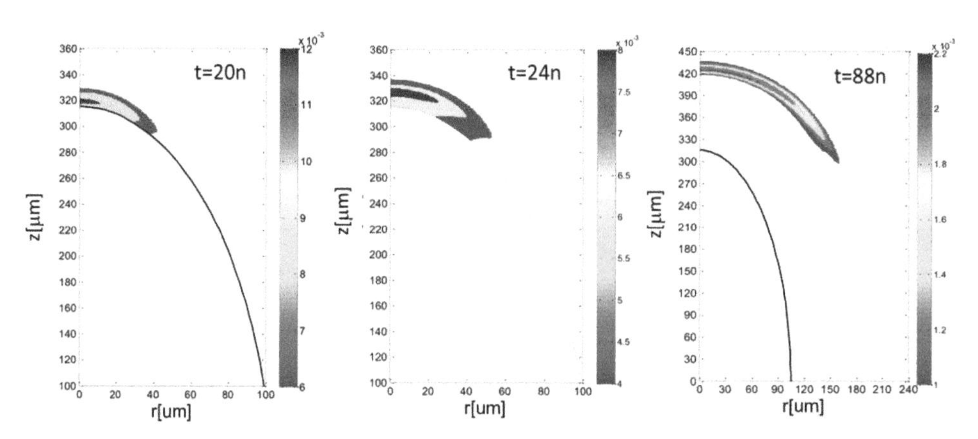


**Figure 5.14.** Two dimensional distributions of the density perturbation $\delta\rho/\rho_0$. The black curve corresponds to the electrode. Reproduced from [7].

It is important to note that the electrostrictive negative pressure depends on the electrode tip radius as $|P_E| \propto E^2 \propto 1/r_{el}^2$, which means that a 30% uncertainty of the tip radius leads to a twofold change of the electrostrictive negative pressure value.

Figures 5.13 and 5.14 show the perturbation of liquid density $\delta\rho/\rho_0$ at different times for $U_0 = 20$ kV The evolution of the density perturbation explains the presence of the bright spots in figure 5.11. The position of the maxima of luminosity in figure 5.11 coincides with the position of the maximum density of the liquid in figure 5.13. Thus, at $t = 88$ ns, the maximum luminance is at a distance of $l \approx 93$ μm from the electrode in the experiment (figure 5.11, fifth image) and the corresponding calculated maximum density is at $l \approx 110$ μm (curve 5 in figure 5.13 and the third plot in figure 5.14).

## 5.6 Initiation of cavitation and nanosecond breakdown in oil on water microdroplets [13]

The dielectric strength of transformer oil depends on its water content [14–18], which may be present in dissolved form, in the form of an emulsion, or droplets. The mechanism of breakdown in oil is associated with the occurrence of vapor bubbles arising in a process of local heating of oil on the impurities [14–18]. Under a nanosecond voltage pulse on the electrodes of a high-voltage oil capacitor or transformer, electrolysis (evaporation of water droplets forming the vapor) does not occur. In this section we will show how water droplets of micron size can stimulate the breakdown of oil at a nanosecond voltage pulse.

Consider a water micro-droplet located in the volume of transformer oil. In this case, the problem of the distribution of the electric field near the droplet is reduced to the field near the dielectric sphere with a dielectric constant of water $\varepsilon_{in} = 81$ in a medium with $\varepsilon_{out} = 2.3$ (see table 3.4) in a uniform external field changing with time. The total pressure in oil, as in the case of water, is the sum of the negative pressure associated with the electrostrictive forces and hydrostatic pressure. Using the formulas for the radial and the azimuthal fields (4.46), we obtain:

$$p_{total} = p + P_E = p - \frac{\varepsilon_0}{2}\frac{(\varepsilon_{out} - 1)(\varepsilon_{out} + 2)}{3\varepsilon_{out}}E^2, \tag{5.17}$$

where

$$E^2 = E_0^2\left(1 + \left(3\cos^2(\theta) + 1\right)\left(\frac{\varepsilon_{out} - \varepsilon_{in}}{2\varepsilon_{out} + \varepsilon_{in}}\right)^2\frac{R^6}{r^6} - \left(5\cos^2(\theta) - 1\right)\frac{R^3}{r^3}\frac{\varepsilon_{out} - \varepsilon_{in}}{2\varepsilon_{out} + \varepsilon_{in}}\right). \tag{5.18}$$

In (5.17) we take into account that oil is a non-polar liquid and $\rho\partial\varepsilon/\partial\rho$ is given by (4.90). $E_0$ in (5.18) is the external field, which generally depends on time, and $R$ is the droplet radius.

Figure 5.15(a) shows the distribution $p_{total}$ (in units $|P_E| = (\varepsilon_{out} - 1)(\varepsilon_{out} + 2)E_0^2/6$) around a microdroplet of water, in a case when the hydrostatic pressure can be neglected. The negative pressure is seven to eight times greater in the vicinity of a water microdroplet's pole than the pressure in the oil and, in accordance with (5.17), the region of the elevated negative pressure grows with increase of the droplet radius.

Figure 5.15(b) shows the corresponding distribution of the electric field along the line passing through the droplet pole (along the $z$-axis). In the vicinity of the water droplet's pole, the electric field in the oil formation exceeds the unperturbed field by more than a factor of 2. That is, if the conditions for cavitation formation and breakdown are not met in pure oil, they occur with large excess in the polar regions of water microdroplets. Thus, when the electric field is turned on quickly, water microdroplets may initiate breakdown in the transformer oil, i.e. determine its dielectric strength at essentially pre-breakdown values of the electric field.

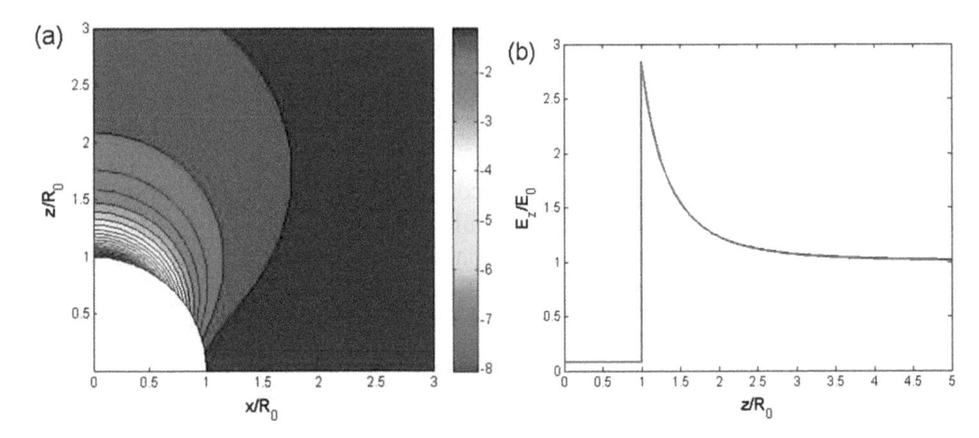


**Figure 5.15.** Plot (a) shows the distribution of the electrostrictive pressure in transformer oil of a water droplet (normalized to $|P_E|$). Plot (b) shows the corresponding electric field along the $z$-axis, normalized to $E_{out}$. Reproduced from [13].

A condition for the front voltage pulse duration $\tau_0$, at which $p \ll |P_E|$ in (5.17), was obtained in chapter 1: the time of oil inflow to the droplet $\tau_p$ should be much longer than the characteristic time $\tau_0$ of change of the electric field. Since in the case of water droplets in oil the droplets act as the electrodes, the droplet radius $R$ should be used in the formula (1.40) instead of $r_{el}$. Consequently the condition under which the hydrostatic pressure may be neglected in (5.17) becomes

$$\tau_0 \ll \tau_p \sim 1.55R/c_{s,oil}, \tag{5.19}$$

where $c_{s,oil} \approx 1400$ m s$^{-1}$ is the sound velocity in oil (see table 3.3). At $R \sim 1$ µm, $\tau_0 \ll \tau_p \approx 1.1$ ns.

We will discuss the dynamics of nanopore expansion through the action of ponderomotive forces in chapter 6, and the cavitation initiation of the nanosecond breakdown in chapter 9.

## 5.7 Qualitative analysis of a drop deformation in the pulsed electric field

We have neglected the deformation of a drop in the electric field. Many scientists have studied the question of water (oil) droplet deformation in strong electric fields, both theoretically and experimentally (e.g. [19–25]). The stretching of droplets along the external field and the transformation to ellipsoidal shape is very important for electric breakdown initiation and for disintegration into smaller droplets [20, 21]. However, the transformation of a spherical droplet to an elongated ellipsoid occurs in a time which is much greater than a nanosecond range [22].

When calculating the droplet deformation, it is necessary to take into account the attached mass of oil and friction forces along with surface tension. These factors slow down the stretching rate of a droplet along the electric field. We will now show that, even without these factors, the required field $E_0$ for a noticeable stretching of a drop during the time of a few nanoseconds is several orders of magnitude greater than the typical value of the breakdown field in oil.

First, we will find the average force necessary for stretching a drop along the direction of the electric field. Since we are interested in time intervals of a few nanoseconds, we can neglect with the hydrostatic pressures $p_{\text{in}}$ and $p_{\text{out}}$ on both sides of the boundary when considering the forces acting on the interface boundary (4.97). Integrating (4.97) by $\pi R_0^2 \sin(\theta)\cos(\theta)\, \mathrm{d}\theta$, where $R_0$ is the radius of the drop, we obtain

$$\begin{aligned} F &= 2\pi R_0^2 \int_0^{\pi/2} F_\Gamma \sin(\theta)\cos(\theta)\mathrm{d}\theta = \pi R_0^2 \varepsilon_0 E_0^2 \left( \frac{3\varepsilon_{\text{out}}}{\varepsilon_{\text{in}} + 2\varepsilon_{\text{out}}} \right)^2 \\ &\quad \left( \int_0^{\pi/2} \left(\varepsilon_{\text{in}}\left(1 - \alpha_{\text{in},E}\right) - \varepsilon_{\text{out}}\left(1 - \alpha_{\text{out},E}\right)\right)\sin^3(\theta)\cos(\theta)\mathrm{d}\theta \right. \\ &\quad \left. + \int_0^{\pi/2} \frac{\varepsilon_{\text{in}}}{\varepsilon_{\text{out}}} \left(\varepsilon_{\text{in}}\left(1 + \alpha_{\text{out},E}\right) - \varepsilon_{\text{out}}\left(1 + \alpha_{\text{in},E}\right)\right)\cos^3(\theta)\sin(\theta)\mathrm{d}\theta \right) \\ &= \frac{\pi R_0^2 \varepsilon_0 E_0^2}{4} \left( \frac{3\varepsilon_{\text{out}}}{\varepsilon_{\text{in}} + 2\varepsilon_{\text{out}}} \right)^2 \left( \varepsilon_{\text{in}}\left(1 - \alpha_{\text{in},E}\right) - \varepsilon_{\text{out}}\left(1 - \alpha_{\text{out},E}\right) \right. \\ &\quad \left. + \frac{\varepsilon_{\text{in}}}{\varepsilon_{\text{out}}} \left(\varepsilon_{\text{in}}\left(1 + \alpha_{\text{out},E}\right) - \varepsilon_{\text{out}}\left(1 + \alpha_{\text{in},E}\right)\right) \right). \end{aligned} \tag{5.20}$$

Taking into account that for water droplets in oil $\varepsilon_{\text{in}} \approx 80$, $\alpha_{\text{in,E}} = 1.5$, $\varepsilon_{\text{out}} = 2.5$, and $\alpha_{\text{out,E}} = (\varepsilon_{\text{out}} - 1)(\varepsilon_{\text{out}} + 2)/3\varepsilon_{\text{out}} = 0.9$, equation (5.20) reduces to

$$F \approx 12\pi R_0^2 \varepsilon_0 E_0^2. \tag{5.21}$$

Equating this force to half the mass of the drop $M = \frac{2}{3}\pi R_0^3 \rho$ multiplied by the acceleration $\mathrm{d}^2\xi/\mathrm{d}t^2$ (figure 5.16) and taking into account the approximate restoring force produced by the surface tension, we obtain an approximate equation of motion which is sufficient for simple estimates:

$$\frac{2}{3}\pi R_0^3 \rho \frac{\mathrm{d}^2\xi}{\mathrm{d}t^2} \approx 12\pi R_0^2 \varepsilon_0 E_0^2 - \pi R_0 \sigma_{0\text{s}}, \tag{5.22}$$

where $\xi$ is the elongation of a drop (figure 5.16), $\pi R_0 \sigma_{0\text{s}}$ is the lower estimate of the surface tension force along the electric field direction that prevents elongation of the drop, and $\sigma_{0\text{s}} = 0.072$ N m$^{-1}$ is the surface tension coefficient of water.

Estimating the acceleration in the left-hand side of equation (5.22) as $\mathrm{d}^2\xi/\mathrm{d}t^2 \sim R_0/t^2$, we obtain

$$E_0 \approx \sqrt{\frac{\rho R_0^2}{18\varepsilon_0 t^2} + \frac{\sigma_{0\text{s}}}{12\varepsilon_0 R_0}}. \tag{5.23}$$

From this it follows that the surface tension forces can be neglected if the radius of the water droplet is

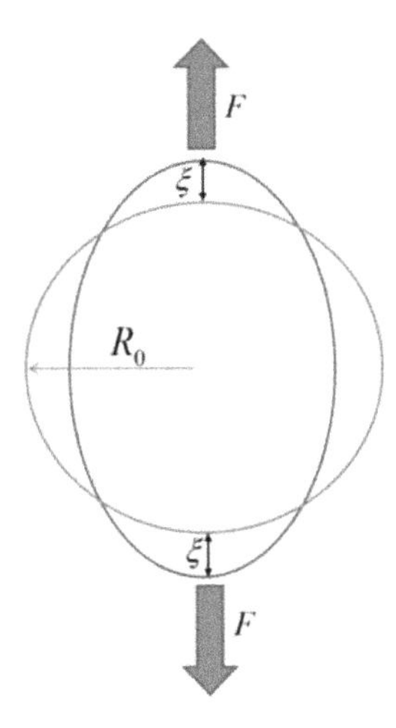


**Figure 5.16.** Stretching of a water droplet due to the forces acting on the water–oil interface boundary.

$$R_0 \gg \left(\frac{3t^2\sigma_{0s}}{2\rho}\right)^{1/3}. \tag{5.24}$$

For a time interval of $t \sim 1$ ns, this omission is valid for $R_0 \gg 50$ nm. If, for example, the radius of the drop is $R_0 = 1$ μm and the pulse duration is $t \approx 1$ ns, the field at which the drop will stretch out by a factor of 2 during the pulse is $E_0 \approx 2.5 \cdot 10^9$ Vm$^{-1}$. Thus, to obtain a noticeable water droplet deformation at a nanosecond time, the external field should be an order of magnitude greater than the critical breakdown field $E_{br} \approx 3{-}5 \cdot 10^8$ Vm$^{-1}$ (see chapter 9). The continuity of the oil around the water droplets will break much earlier due to the ponderomotive electrostrictive tensions. So at sub-nanosecond and nanosecond time intervals one can neglect the influence of an electric field on a droplet's shape.

## References

[1] Shneider M N and Pekker M 2013 Dielectric fluid in inhomogeneous pulsed electric field *Phys. Rev.* E **87** 043004

[2] Landau L D and Lifshitz E M 1987 *Fluid Mechanics* 2nd edn (Oxford: Pergamon Press)

[3] Nimtz G and Weiss W 1987 Relaxation time and viscosity of water near hydrophilic surfaces *Z. Phys.* B **67** 483

[4] Morse P M and Feshbach H 1953 *Methods of Theoretical Physics. Part I* (New York: McGraw-Hill)

[5] Anderson D A, Tannehill J C and Pletcher R H 1984 *Computational Fluid Mechanics and Heat Transfer* (New York: Hemisphere)

[6] Shneider M N, Pekker M and Fridman A 2012 *IEEE Trans. Dielectr. Electr. Insul.* **19** 1579

[7] Pekker M, Seepersad Y, Shneider M N, Fridman A and Dobrynin D 2013 Initiation stage of nanosecond breakdown in liquid *J. Phys. D: Appl. Phys.* **47** 025502

[8] Dobrynin D, Seepersad Y, Pekker M, Shneider M N, Friedman G and Fridman A 2013 Non-equilibrium nanosecond-pulsed plasma generation in the liquid phase (water, PDMS)

without bubbles: fast imaging, spectroscopy and leader-type model *J. Phys. D: Appl. Phys.* **46** 105201
[9] Seepersad Y, Pekker M, Shneider M N, Dobrynin D and Fridman A 2013 On the electrostrictive mechanism of nanosecond-pulsed breakdown in liquid phase *J. Phys. D: Appl. Phys.* **46** 162001
[10] Seepersad Y, Pekker M, Shneider M N, Dobrynin D and Fridman A 2013 Investigation of positive and negative modes of nanosecond pulsed discharge in water and electrostriction model of initiation *J. Phys. D: Appl. Phys.* **46** 355201
[11] Shneider M N and Pekker M 2016 Rayleigh scattering on the cavitation region emerging in liquids *Opt. Lett.* **41** 1090
[12] Settles C S 2006 Schlieren and shadowgraph techniques *Visualizing Phenomena in Transparent Media* 2nd edn (Berlin: Springer)
[13] Pekker M and Shneider M N 2015 Pre-breakdown cavitation nanopores in the dielectric fluid in the inhomogeneous, pulsed electric fields *J. Phys. D: Appl. Phys.* **48** 424009
[14] Koch M, Fischer M and Tenbohlen S 2007 The breakdown voltage of insulation oil under the influences of humidity, acidity, particles and pressure *Int. Conf. APTADM (Wroclaw, Poland, 2007)*
[15] Griffin P J 1995 Water in transformers—so what! *National Grid Conf. on Condition Monitoring in High Voltage Substations (Dorling, 1995)*
[16] Gockenbach E and Borsi H 2002 Performance and new application of ester liquids *Proc. 2002 IEEE 14th Int Conf. on Dielectric Liquids* (*Graz, Austria, 2002*)
[17] Gradnik T, Koncan-Gradnik M, Petric N and Muc N 2011 Experimental evaluation of water content determination in transformer oil by moisture sensor *IEEE Int Conf. Dielectric Liquids* (*Trondheim, Norway, 2011*)
[18] Feely C 2006 Transformer moisture monitoring and dehydration—Powercor experience *Tech.Con. Asia-Pacific* (*Sydney, Australia, 2006*)
[19] Zeleny J 1916 On the conditions of instability of electrified drops, with applications to the electrical discharge from liquid points *Proc. Camb. Phil. Soc.* **18** 71
[20] Taylor G 1964 Disintegration of water drops in electric field *Proc. R. Soc.* A **280** 383
[21] Sherwood J D 1988 Breakup of fluid droplets in electric and magnetic fields *J. Fluid Mech.* **188** 133
[22] Zhang J, Zahn J D and Lin H 2013 Transient solution for droplet deformation under electric fields *Phys. Rev.* E **87** 043008
[23] Miksis M J 1981 Shape of a drop in an electric field *Phys. Fluids* **24** 1967
[24] Grigor'ev A I 2000 Electric dispersion of fluid under the oscillatory instability of its free surface *Tech. Phys.* **45** 543
[25] Garton C G and Krasucki Z 1964 Bubbles in insulating liquids: stability in an electric field *Proc. R. Soc.* A **280** 211

**M N Shneider and M Pekker**

# Chapter 6

# Cavitation in inhomogeneous pulsed electric fields

*In chapter 5, we found conditions under which the tension in homogeneous polar liquids (e.g. water) or in non-polar liquids (e.g. oil) leads to the formation of ruptures (nanopores) caused by the electrostrictive ponderomotive forces. But we did not yet consider the subsequent time evolution. In this chapter, we will address two important problems: whether nanopores collapse under the action of surface tension forces or, in contrast, increase, and whether negative pressure exceeds the threshold of cavitation. Specifically, what is the highest density of bubbles that can emerge in a liquid in the vicinity of the pin electrode with the applied voltage pulse?*

## 6.1 Ponderomotive forces in the vicinity of a nanopore

We will now find the value of the square of the electric field in the area where a bubble emerges. Substituting $\varepsilon_{\text{in}} = 1$, $p_{\text{in}} = 0$, $\alpha_{\text{in,E}} = 0$, $\varepsilon_{\text{out}} = \varepsilon$, $p_{\text{out}} = p_{\text{h}}$, and $\alpha_{\text{out,E}} = \alpha_{\text{E}}$ into (4.45), we obtain

$$E_{\text{out}}^2 = E_0^2 + E_0^2\left(\left(3\cos^2(\theta) + 1\right)\left(\frac{(\varepsilon - 1)}{2\varepsilon + 1}\right)^2\frac{R^6}{r^6} - \left(5\cos^2(\theta) - 1\right)\frac{R^3}{r^3}\frac{(\varepsilon - 1)}{2\varepsilon + 1}\right), \quad (6.1)$$

in which $E_0$ is the value of the unperturbed electric field at the bubble's location, and the second term on the right-hand side of (6.1) is the square of the electric field perturbation in the liquid caused by the bubble. Figure 6.1 shows the radial dependence of the square of the electric field for several values of the azimuthal angle for water ($\varepsilon = 81$, $\alpha_{\text{E}} = 1.5$) and for transformer oil ($\varepsilon = 2.25$, $\alpha_{\text{E}} \approx 0.8$).

doi:10.1088/978-0-7503-1245-5ch6

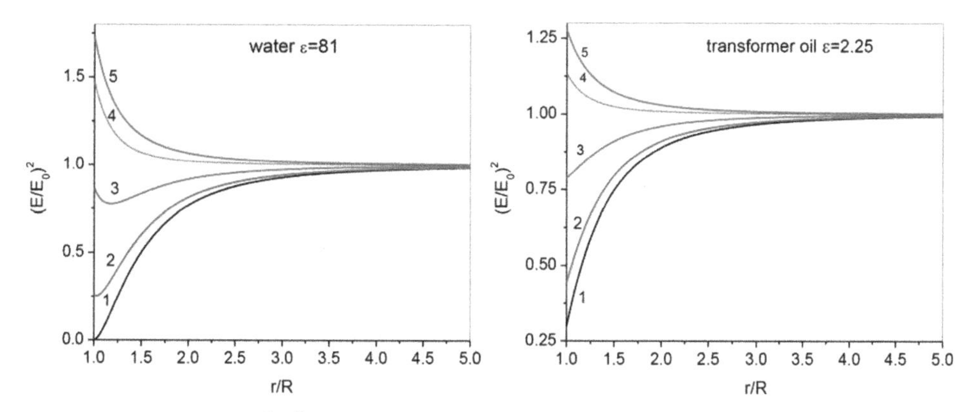


**Figure 6.1.** Dependence of $E^2/E_0^2$ on $r/R$ at different angles $\theta$ for water (left) and transformer oil (right). Curve 1 corresponds to $\theta = 0$, curve 2 to $\theta = \pi/8$, curve 3 to $\theta = \pi/4$, curve 4 to $\theta = 3\pi/8$, and curve 5 to $\theta = \pi/2$.

Since dielectric fluid always moves toward the highest electric field, according to figure 6.1, the liquid flows from the pore at the poles (curves 1, 2, and 3), and to the pore at the equator (curves 4 and 5).

In accordance with (6.1), the negative pressure acting on the liquid in the pore's vicinity is

$$P_- = -\frac{\alpha_E \varepsilon_0 \varepsilon E_0^2}{2} - \frac{\alpha_E \varepsilon_0 \varepsilon E_0^2}{2}\left(\left(3\cos^2(\theta) + 1\right)\left(\frac{(\varepsilon - 1)}{2\varepsilon + 1}\right)^2 \frac{R^6}{r^6} - \left(5\cos^2(\theta) - 1\right)\frac{R^3}{r^3}\frac{(\varepsilon - 1)}{2\varepsilon + 1}\right). \tag{6.2}$$

Consequently, the components of the ponderomotive electrostrictive forces near the pore are

$$F_r = -\frac{\partial P_-}{\partial r} \approx -\frac{3P_E}{r}\frac{\varepsilon - 1}{2\varepsilon + 1}\frac{R^3}{r^3}\left(\left(5\cos^2(\theta) - 1\right) - 2\left(3\cos^2(\theta) + 1\right)\cdot\frac{\varepsilon - 1}{2\varepsilon + 1}\frac{R^3}{r^3}\right) \tag{6.3}$$

$$F_\theta = -\frac{1}{r\sin\theta}\frac{\partial P_-}{\partial\theta} \approx \frac{2P_E}{r}\frac{\varepsilon - 1}{2\varepsilon + 1}\frac{R^3\cos(\theta)}{r^3}\cdot\left(3\cdot\left(\frac{\varepsilon - 1}{2\varepsilon + 1}\right)\frac{R^3}{r^3} - 5\right), \tag{6.4}$$

where $P_E = -\alpha_E \varepsilon_0 \varepsilon E_0^2/2$. We neglect the gradient of $E_0^2$ in these equations, because the scale of the variation of $E_0$ is of order of the radius of curvature of the tip of the electrode, which is much greater than the size of the cavitation bubble in the early stages of its growth.

Figures 6.2 and 6.3 show the radial dependence and two-dimensional contour plot of normalized $F_r$ for water and oil, respectively.

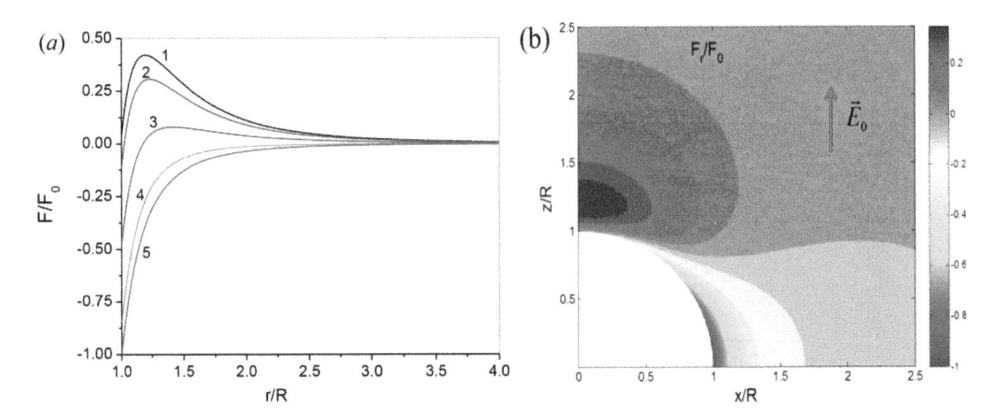


**Figure 6.2.** Results for water. Radial dependence (a) and two-dimensional contour plot (b) of the normalized volumetric force $F_r(r, \theta)$ in the vicinity of the pore at different angles $\theta$. In (a), curve 1 corresponds to $\theta = 0$, curve 2 to $\theta = \pi/8$, curve 3 to $\theta = \pi/4$, curve 4 to $\theta = 3\pi/8$, and curve 5 to $\theta = \pi/2$. $F_0 = |F_r(R, \pi/2)|$. Panel (a) reproduced from [1]. Copyright 2015 AIP Publishing LLC.

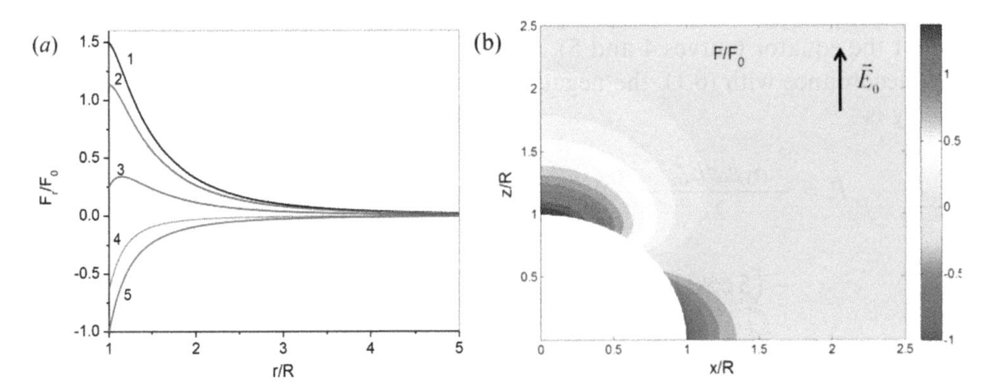


**Figure 6.3.** The same as figure 6.2 but for oil.

Ponderomotive forces, as we have repeatedly mentioned, cause fluid to move toward the greater electric field; that is, to where the hydrostatic pressure forms that compensates negative pressure. Since, for the time of a sub-nanosecond or nanosecond pulse the leaking of the liquid to the electrode is very small (see estimate (1.40) and chapter 5), the hydrostatic pressure change during this time can be neglected. The initial unperturbed pressure $p_0$, which is close to atmospheric pressure ($p_0 \approx 0.1$ MPa), can also be neglected in comparison with the negative pressure at which the liquid loses its homogeneity, the absolute value of which is several tens of MPa.

We estimate the characteristic time $\tau_\text{p}$ of fluid leakage (inflow) to the nanopore, which is associated with the ponderomotive force (6.3), on the basis of the linearized one-dimensional equation of motion:

$$\rho \frac{\partial u}{\partial t} = F_r \sim \frac{P_\text{E}}{R}. \tag{6.5}$$

Substituting $\partial u/\partial t \sim R/t^2$, we obtain the estimate $\tau_p \sim R\sqrt{\rho/|P_E|}$. For example, for the pore with radius $R = 2$ nm in water, the stretching forces exceed the Laplace pressure at $|P_E| \geqslant 250$ MPa ($E_0 \geqslant 8 \cdot 10^8$ V m$^{-1}$), so $\tau_p \sim 4$ ps. Assuming that the characteristic time of the electrostrictive pore expansion is smaller than the time for the fluid leakage (inflow), the negative pressure associated with the electric field perturbation will be completely compensated by hydrostatic pressure. Therefore,

$$p(r, \theta) = -P_E\left(\left(3\cos^2(\theta) + 1\right) \cdot \left(\frac{\varepsilon - 1}{2\varepsilon + 1}\right)^2 \frac{R^6}{r^6} - \left(5\cos^2(\theta) - 1\right) \cdot \frac{R^3}{r^3} \cdot \frac{\varepsilon - 1}{2\varepsilon + 1}\right). \tag{6.6}$$

## 6.2 Nucleation in inhomogeneous pulsed electric fields

The expression for the free energy of a bubble (chapter 2, section 2.5) has the form:

$$W(R_b) = -2\pi\int_0^{R_b}\int_0^{\pi} |P_-|\sin(\theta)r^2 \mathrm{d}r\mathrm{d}\theta + 8\pi\int_0^{R_b} r^2\frac{\sigma_s}{r}\mathrm{d}r. \tag{6.7}$$

The surface tension coefficient $\sigma_s$ is defined by formula (2.10). In section 2.5 we assumed that the negative pressure $P_-$ is constant and does not depend on the azimuthal angle $\theta$. However, this is not true for negative pressure created by ponderomotive forces, because the electric field is discontinuous at the interface of the pore. The force per unit surface area on the nanopore was derived in chapter 4 in the approximation of an infinitely thin interface layer at the pore's boundary. Substituting into (4.99) $\varepsilon_{in} = 1$, $p_{in} = 0$, $\alpha_{in,E} = 0$, $\varepsilon_{out} = \varepsilon$, $p_{out} = p_h$, and $\alpha_{out,E} = \alpha_E$, we obtain

$$\begin{aligned} F_r^{sphere} = P_E\Bigg(&\left(3\cos^2(\theta) + 1\right)\left(\frac{\varepsilon - 1}{2\varepsilon + 1}\right)^2 - \left(5\cos^2(\theta) - 1\right)\frac{\varepsilon - 1}{2\varepsilon + 1}\Bigg) \\ &- P_E\left(\frac{3\varepsilon}{1 + 2\varepsilon}\right)^2\Bigg(\left(\sin^2(\theta) + \frac{1}{\varepsilon^2}\cos^2(\theta)\right) \\ &- \frac{(\varepsilon - 1)}{\alpha_E\varepsilon}\left(\sin^2(\theta) + \frac{1}{\varepsilon}\cos^2(\theta)\right)\Bigg) \end{aligned} \tag{6.8}$$

The first expression in parentheses in (6.8) corresponds to the hydrostatic pressure $p_h(R, \theta)$ (6.8) with a negative sign, because it acts on the outside surface of the bubble (figure 4.17).

Figure 6.4 shows how the force acting on a unit area of the pore (6.9) for water and transformer oil varies with azimuthal angle.

When comparing the curves associated with prompt turning of the electric field in figure 6.4 to those in figures 4.18 and 4.19, one sees that while the stretching forces acting on a bubble are maximal near the pole in figure 6.4, in figures 4.18 and 4.19

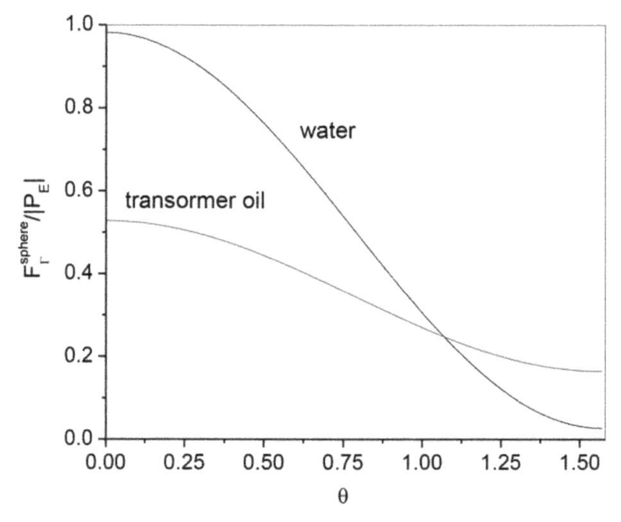


**Figure 6.4.** Dependence of the force acting on unit area of a spherical pore on the azimuthal angle in water and transformer oil.

they are maximal near the equator. Specifically, the bubble in figure 6.4 is stretched along the electric field, while those in figures 4.18 and 4.19 are flattened out. This effect is related to the liquid inflow to the equator and its outflow from the pole during the bubble emergence and development. The fluid flows away from the pore for azimuthal angles of

$$\theta < \arccos\left(\sqrt{\frac{3\varepsilon}{7\varepsilon + 8}}\right), \tag{6.9}$$

while at greater $\theta$, it flows toward the pore. Inequality (6.9) is obtained from (6.6) at $r = R$.

Since the tensile forces are reduced at the equator (figure 6.4), the stretching of a spherical pore along the direction of the electric field is accompanied by its compression at the equator due to the Laplace pressure:

$$\begin{aligned} F_{\Gamma,\text{total}}^{\text{sphere}} &= P_{\text{E}}\left(\left(3\cos^2(\theta) + 1\right)\left(\frac{\varepsilon - 1}{2\varepsilon + 1}\right)^2 - \left(5\cos^2(\theta) - 1\right)\cdot\frac{\varepsilon - 1}{2\varepsilon + 1}\right) \\ &- P_{\text{E}}\left(\frac{3\varepsilon}{1 + 2\varepsilon}\right)^2\left(\left(\sin^2(\theta) + \frac{1}{\varepsilon^2}\cos^2(\theta)\right)\right. \\ &- \left.\frac{(\varepsilon - 1)}{\alpha_{\text{E}}\varepsilon}\left(\sin^2(\theta) + \frac{1}{\varepsilon}\cos^2(\theta)\right)\right) - \frac{2\sigma_{\text{s}}}{R}. \end{aligned} \tag{6.10}$$

In other words, the nanopore's elongation occurs without any significant change in its volume. Figure 6.5 shows a qualitative picture of the changing shape of the nanopore when the liquid inflow to the electrode can be neglected.

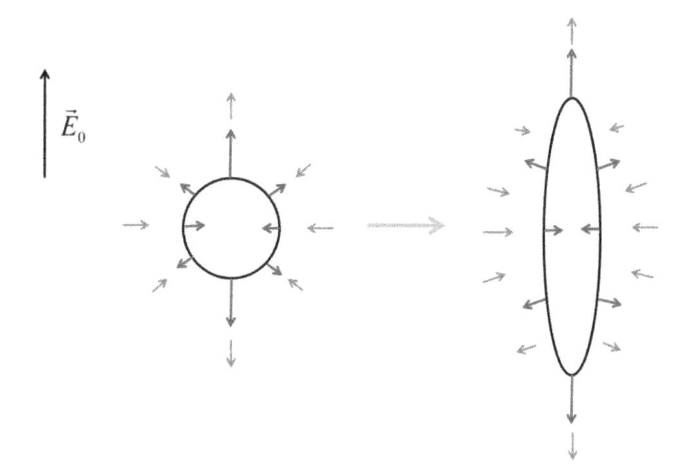


**Figure 6.5.** Qualitative picture of a micropore's shape during formation. $\vec{E}_0$ is the unperturbed external field produced by the voltage applied to the electrode. The red arrows indicate the forces acting on the liquid–pore boundary and the blue arrows demonstrate the directions of the fluid motion in the outer region of the pore induced by the electrostrictive volumetric force.

The above considerations are mostly qualitative and valid only for the fluid outside a pore assumed as an elastic medium, i.e. for a very small deformation of the micropore and fluid in its vicinity. In order to determine the exact shape of the expanding pore we need to consider the following factors in the vicinity of the pore: boundary conditions for the ponderomotive forces, tension forces, the hydrostatic pressure, and the appearance of new microscopic discontinuities (nanopores) in the fluid in the region of large tensile stresses.

Substituting $P_- = F_\Gamma^{\text{sphere}}$ in (6.7), where $F_\Gamma^{\text{sphere}}$ is determined by (6.8), and the surface tension $\sigma_s$ (2.10) after integration we obtain

$$W(R_b) = -\frac{\alpha_E \varepsilon_0 \varepsilon k_E E_0^2}{2}\frac{4\pi R_b^3}{3} + 4\pi R_b^2 \sigma_{0s} - 4\pi \sigma_{0s} \delta_b^2 \ln\left(1 + \frac{R_b^2}{\delta_b^2}\right), \tag{6.11}$$

where

$$k_E = 2\left(\frac{\varepsilon - 1}{2\varepsilon + 1}\right)^2 - \frac{2}{3}\frac{\varepsilon - 1}{2\varepsilon + 1} + \left(\frac{3\varepsilon}{1 + 2\varepsilon}\right)^2\left(\frac{2\varepsilon^2 + 1}{3\varepsilon^2} - \frac{(\varepsilon - 1)(2\varepsilon + 1)}{3\alpha_E \varepsilon^2}\right). \tag{6.12}$$

For water, $\varepsilon = 81$, $\alpha_E = 1.5$, and $k_E = 0.346$; for oil, $\varepsilon = 2.25$, $\alpha_E \approx 0.8$, and $k_E = 0.2852$; and for liquid helium, $\varepsilon = 1.057$, $\alpha_E \approx 0.057$, and $k_E = 0.013$.

Equation (6.11) allows one to determine the magnitude of the pulsed electric field $E_{cr}(r, t)$ that corresponds to the negative pressure $P_{cr}$ at which cavitation begins. Since

$$|P_{cr}| = \frac{\alpha_E \varepsilon_0 \varepsilon k_E E_{cr}^2}{2}, \tag{6.13}$$

by substituting (6.13) into (2.13) and (2.15), we obtain expressions for $R_{cr}$ and $W_{cr}$ depending on the electric field. If, for example, we know from the experiment that cavitation starts at the negative pressure $|P_{cr}| = 30$ MPa, then from (6.12) we can determine the corresponding minimum electric field and, respectively, the voltage on the electrode at which cavitation begins. On the other hand, if we know the field at which cavitation begins in the liquid from the experiment, then we can figure out the corresponding critical negative pressure from (6.13).

## 6.3 Expansion of nanopores in an inhomogeneous pulsed electric field

We will now consider the problem of a spherical pore's expansion under the influences of the electrostrictive ponderomotive forces. Substituting

$$P_{\text{in}} - P_{\text{out}} = 2\pi \int_0^{\pi} F_{\Gamma}^{\text{sphere}} \sin(\theta) \mathrm{d}\theta = 2\pi \alpha_{\mathrm{E}} \varepsilon_0 \varepsilon k_{\mathrm{E}} E_0^2, \tag{6.14}$$

and surface tension (2.1) into (1.7), we obtain

$$\frac{\mathrm{d}}{\mathrm{d}t}\left(R^3\left(\frac{\mathrm{d}R}{\mathrm{d}t}\right)^2\right) = \frac{4}{\rho} R \frac{\mathrm{d}R}{\mathrm{d}t}\left(\frac{\alpha_{\mathrm{E}} \varepsilon_0 \varepsilon k_{\mathrm{E}} E_0^2 R}{4} - \frac{\sigma_{0\mathrm{s}}}{1 + (\delta_{\mathrm{b}}/R)^2}\right). \tag{6.15}$$

In (6.14) and (6.15) $E_0$ is the electric field in a point of the nanopore location at time $t$. While it is the saturated vapor pressure that prevents a Rayleigh bubble from collapsing, the electric field difference on the liquid–vapor interface plays this role for a bubble in an electric field.

Employing new variables

$$\begin{matrix} R = R_0 x^{2/5} \\ t = t_0 \tau \end{matrix}, \tag{6.16}$$

equation (6.15) takes the form

$$\frac{\mathrm{d}^2 x}{\mathrm{d}\tau^2} = \frac{5 t_0^2 \sigma_{0\mathrm{s}}}{R_0^3 \rho} x^{1/5}\left(\frac{R_0}{2\sigma_0}|P_-| - \frac{x^{2/5}}{x^{4/5} + (\delta_{\mathrm{b}}/R_0)^2}\right), \tag{6.17}$$

where, $P_- = -\alpha_{\mathrm{E}} \varepsilon_0 \varepsilon k_{\mathrm{E}} E_0^2(\tau, r)/2$ is the negative pressure acting on the nanopore located at point $r$ at a dimensionless time $\tau$.

Parameters $\delta_{\mathrm{b}}$ and $R_0$ are related to the critical pressure $P_{\mathrm{cr}} = -30$ MPa. Since $\delta_{\mathrm{b}}$ and $R_{\mathrm{cr}}$ are coupled by (2.13), we set $\delta_{\mathrm{b}} = 2.4$ nm and $R_0 = R_{\mathrm{cr}} = 2.5$ nm. We consider the event of an instantaneous field turning on. Figure 6.6 shows the time evolution of the nanopore radius and its expansion velocity for water and oil. An absolute value of the negative pressure of 100 MPa is related to $E_0 = 7.3 \cdot 10^8$ V m$^{-1}$ in water and $E_0 = 6.3 \cdot 10^9$ V m$^{-1}$ in oil. A higher expansion velocity in oil can be explained by a lower surface tension coefficient.

From (6.15) follows the asymptotic expansion of the pore, when its radius is so large that the Laplace pressure may be neglected. Indeed, from (6.15) we obtain

$$\frac{\mathrm{d}}{\mathrm{d}t}\left(R^3\left(\frac{\mathrm{d}R}{\mathrm{d}t}\right)^2\right) = 3R^2\left(\frac{\mathrm{d}R}{\mathrm{d}t}\right)^3 + 2R^3 \frac{\mathrm{d}R}{\mathrm{d}t}\frac{\mathrm{d}^2 R}{\mathrm{d}t^2} = \frac{2}{\rho} R^2 \frac{\mathrm{d}R}{\mathrm{d}t}|P_-|. \tag{6.18}$$

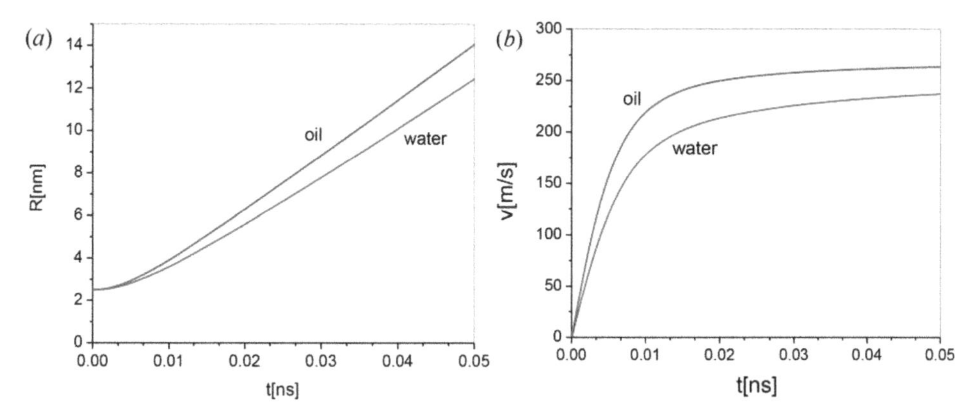


**Figure 6.6.** Expansion of nanopores in water and transformer oil at a constant negative pressure of $P_{max} = |P_{-}| = 100$ MPa. Plot (a) shows the pore's radius and plot (b) its expansion velocity as functions of time.

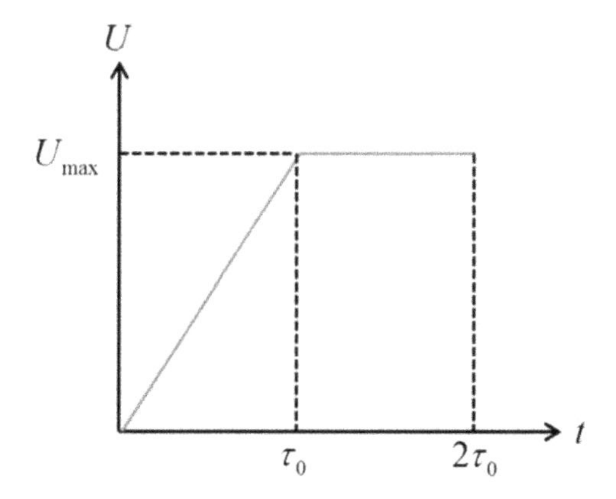


**Figure 6.7.** Time-dependence of the voltage pulse on a spherical electrode.

Since the pore's asymptotic expansion velocity is a constant, from (6.18) follows

$$\left(\frac{\mathrm{d}R}{\mathrm{d}t}\right)_{\mathrm{asym}} = \sqrt{\frac{2P_{\mathrm{max}}}{3\rho}}. \tag{6.19}$$

At $P_{max} = |P_{-}| = 100$ MPa, $(\frac{\mathrm{d}R}{\mathrm{d}t})_{\mathrm{asym}} = 258\ \mathrm{m\ s^{-1}}$ and $272\ \mathrm{m\ s^{-1}}$ for water and oil, respectively. The nanopore expansion rate is substantially subsonic in a liquid, so shock waves do not form.

Consider now the case of a finite-time switching of the electric field. For simplicity, we assume a spherical electrode with a voltage pulse dependence on time shown in figure 6.7.

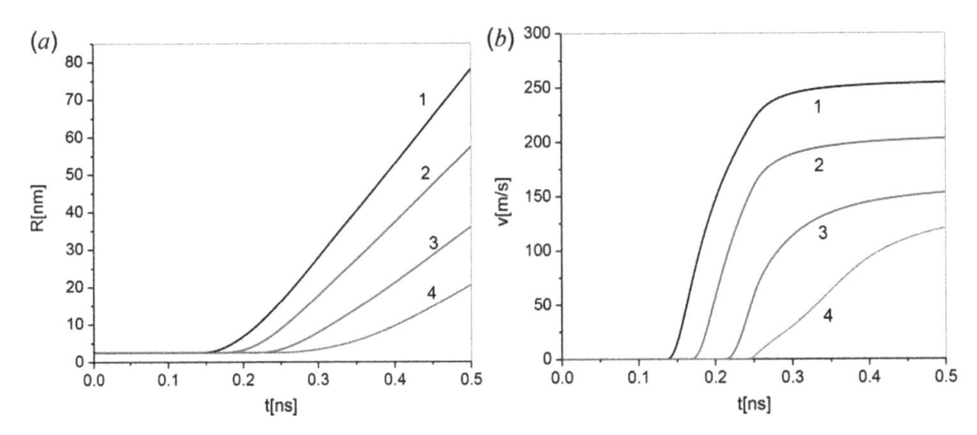


**Figure 6.8.** Nanopore development in water at a time-varying voltage on a spherical electrode shown in figure 6.7 at $\tau_0 = 0.25$ ns and $P_{\max}(r_{\mathrm{el}}, \tau_0) = 100$ MPa. Plot (*a*) shows the radii of nanopores and (*b*) shows the rates of their expansion. Curve 1 corresponds to $r_{\mathrm{el}}/r = 1$, curve 2 to $r_{\mathrm{el}}/r = 0.9$, curve 3 to $r_{\mathrm{el}}/r = 0.8$, and curve 4 to $r_{\mathrm{el}}/r = 0.75$.

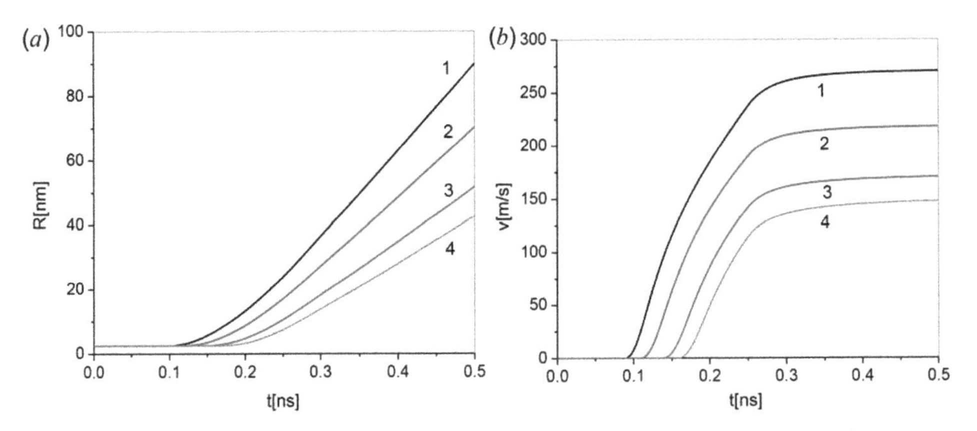


**Figure 6.9.** Development of nanopores in oil for the same parameters as in figure 6.8.

For a given voltage $U(t)$ on a spherical electrode, the negative pressure is

$$|P_{-}(r, t)| = \frac{\alpha_{\mathrm{E}}\varepsilon_0\varepsilon k_{\mathrm{E}} U(t)^2}{2r_{\mathrm{el}}^2}\frac{r_{\mathrm{el}}^4}{r^4}. \tag{6.20}$$

Figures 6.8 and 6.9 show the time evolution of the nanopore radius and its expansion velocity for water and oil at the applied voltage pulse shown in figure 6.7.

## 6.4 Concluding remarks

If we know the electric field at which cavitation begins from the experiment with the nanosecond voltage pulse applied to the needle-like electrode, then from (6.13) we can figure out the corresponding critical negative pressure, which can be measured using the standard methods mentioned in chapter 2.

1. The results that were obtained for the spherical nanopores are only qualitative because, as indicated above, the nanopore stretches along the direction of the external field.
2. The exact calculation of the size and shape of the growing nanopores requires consideration of all the factors listed above.

## Reference

[1] Shneider M N and Pekker M 2015 Pre-breakdown processes in a dielectric fluid in inhomogeneous pulsed electric fields *J. Appl. Phys.* **117** 224902

**M N Shneider and M Pekker**

# Chapter 7

## Liquid helium in a non-uniform pulsed electric field

*In this chapter, remaining within the concept of a conventional continuous liquid, we show that the description of cavitation produced by a pulsed non-uniform electric field in liquid helium is identical to that in a conventional dielectric liquid (chapter 5), regardless of whether the helium is in the normal or superfluid phase. The possibility of observing nucleation and cavitation in liquid helium in a non-uniform pulsed electric field is discussed.*

As was noted in the introductory chapter, cavitation may occur in almost all liquids. To develop cavitation, one must create a sufficiently strong negative pressure (stretching tension) in the liquid. A way to create the necessary negative pressure in dielectric fluids is by applying a pulsed inhomogeneous electric field. In chapters 1, 2, 5, and 6 we reviewed effects related to the breach of continuity, nucleation, and expansion of cavitation bubbles in polar and non-polar liquid dielectrics at nanosecond pulsed inhomogeneous electric fields.

One of the most interesting substances in which cavitation has also been detected is liquid helium, because it can be in the normal or superfluid phases, depending on the temperature. Typically, cavitation in liquid helium is induced by acoustic waves in the volume or by their reflection from the interface surface [1–6].

One might say that the work of Lifshitz and Kaganov [7] stimulated theoretical physicists' interest in the problem of cavitation in liquid helium. They first noticed that at a certain negative pressure liquid helium converts into a metastable state. It is a condition under which helium can be in both the liquid and gas states with the same thermodynamic parameters. Many theoretical and computational studies have investigated the metastable states of liquid helium-4 and helium-3 at negative pressure [8–11].

In this chapter, remaining within the concept of a conventional continuous liquid, we will show that the description of cavitation produced by a pulsed non-uniform

doi:10.1088/978-0-7503-1245-5ch7

electric field in liquid helium is identical to that in a conventional dielectric liquid, regardless of whether the helium is in the normal or superfluid phase.

The ability to initiate cavitation in liquid helium by a pulsed inhomogeneous electric field is very important, because it allows one to determine when and where the cavitation begins quite accurately from optical measurements [1, 6]. With it, one can observe the Rayleigh scattering [12] (see chapter 8) to measure dielectric permittivity, the coefficient of surface tension, and the level of critical negative pressure in liquid helium, depending on its temperature, density, and phase state.

## 7.1 Dynamics of liquid helium in a non-uniform pulsed electric field

Three aspects determine the occurrence and development of cavitation in liquid helium-3 and -4, as well as in water, or any other dielectric liquid. The first is the negative pressure (tensile tension) under which the fluid breaks (i.e. under which the conditions of continuity are violated). The second is the probability that a pore (bubble) will form with a radius at which the negative pressure is greater than the surface pressure tension (Laplace pressure). The third is the expansion dynamics of the bubble. We presented all these aspects in previous chapters with the examples of water and oil. As shown in chapter 6 [13, 14], pore expansion primarily occurs due to the jump of the electric field at the vacuum–liquid interface (or, more precisely, at the saturated vapor–liquid interface), which serves as a 'piston'.

In the first part of this chapter, we will show that the conditions for development of the discontinuity in liquid helium-3 and helium-4 in an inhomogeneous electric field are essentially the same as in an ordinary polar or non-polar liquid dielectric. In the second part, we will discuss the issues of nucleation in liquid helium.

### 7.1.1 Conditions for the discontinuity formation in helium-3 and -4 in an inhomogeneous pulsed electric field

The equations for conservation of mass, momentum, and energy for the two-component liquid helium, neglecting viscosity, are [15]:

$$\frac{\partial\rho}{\partial t} + \vec{\nabla}\cdot\left(\rho_{\rm n}\vec{u}_{\rm n} + \rho_{\rm s}\vec{u}_{\rm s}\right) = 0, \tag{7.1}$$

$$\frac{\partial}{\partial t}\left(\rho_{\rm n}\vec{u}_{\rm n} + \rho_{\rm s}\vec{u}_{\rm s}\right) + \vec{\nabla}\cdot\left(\rho_{\rm n}\vec{u}_{\rm n}\vec{u}_{\rm n} + \rho_{\rm s}\vec{u}_{\rm s}\vec{u}_{\rm s}\right) = -\vec{\nabla}p + \frac{\varepsilon_0\rho}{2}\vec{\nabla}\left(E^2\frac{\partial\varepsilon}{\partial\rho}\right), \tag{7.2}$$

$$\begin{aligned}\frac{\partial}{\partial t}\left(\frac{1}{2}\rho_{\rm n}u_{\rm n}^2 + \frac{1}{2}\rho_{\rm s}u_{\rm s}^2 + \rho e_{\rm i}\right) = -\vec{\nabla}\Bigg\{&\frac{1}{2}\rho_{\rm n}u_{\rm n}^2\vec{u}_{\rm n} + \frac{1}{2}\rho_{\rm s}u_{\rm s}^2\vec{u}_{\rm s} + (\rho e_{\rm i} + p)\vec{u}\\ &+ \rho(\vec{u}_{\rm n} - \vec{u})\left(Ts + \frac{\rho_{\rm n}}{2\rho}(\vec{u}_{\rm n} - \vec{u}_{\rm s})^2\right)\Bigg\} + \rho\vec{u}\cdot\vec{\nabla}\left(\frac{\varepsilon_0}{2}E^2\frac{\partial\varepsilon}{\partial\rho}\right)\end{aligned} \tag{7.3}$$

Here $\rho_{\rm n}$, $\rho_{\rm s}$, $\vec{u}_{\rm n}$, and $\vec{u}_{\rm s}$ are the densities and velocities of the normal and superfluid liquids, $\rho = \rho_{\rm n} + \rho_{\rm s}$ is the total density, $p$ is the hydrostatic pressure, $e_{\rm i}$ is the intrinsic internal energy per unit mass, (formulas for $\rho_{\rm n}$ and $e_{\rm i}$ are given in [16, 17]),

$\vec{u} = (\rho_n \vec{u}_n + \rho_s \vec{u}_s)/\rho$ is the local center-of-mass velocity, $\varepsilon_0$ is the permittivity of free space, $\varepsilon \approx 1.057$ is the dielectric constant of liquid helium, $s$ is the entropy per unit mass, and $T$ is the temperature. Since liquid helium is a non-polar liquid,

$$\frac{\partial \varepsilon}{\partial \rho} = \frac{(\varepsilon - 1)(\varepsilon + 2)}{3\rho} \approx \frac{(\varepsilon - 1)}{\rho} \approx \frac{0.057}{\rho}. \tag{7.4}$$

We will show below that ruptures in liquid helium occur at $|\rho - \rho_0|/\rho_0 \sim 0.05$, where $\rho_0$ is the density of the liquid without an electric field. Therefore, we can use the following expression for the ponderomotive term in (7.2) and (7.3):

$$\frac{\varepsilon_0 \rho}{2} \vec{\nabla}\left( E^2 \frac{\partial \varepsilon}{\partial \rho} \right) \approx \varepsilon_0 \frac{(\varepsilon - 1)(\varepsilon + 2)}{6} \vec{\nabla} E^2 \approx \frac{(\varepsilon - 1)\varepsilon_0 \vec{\nabla} E^2}{2}. \tag{7.5}$$

The system of equations describing the motion of single-component liquid helium (for example, in the normal state) in a nanosecond pulsed electric field has the form

$$\frac{\partial \rho}{\partial t} + \nabla(\rho \vec{u}) = 0 \tag{7.6}$$

$$\rho\left( \frac{\partial \vec{u}}{\partial t} + (\vec{u} \cdot \nabla)\vec{u} \right) = -\nabla p_{\text{total}} \tag{7.7}$$

$$p_{\text{total}} = p + P_E = p - \beta_E E^2 \tag{7.8}$$

where $\beta_E = \frac{1}{2}(\varepsilon - 1)\varepsilon_0$.

Equations (7.6)–(7.8) coincide with the equations of water dynamics in a pulsed non-uniform electric field, (5.1) and (5.4), combined with another equation of state, $p = p(\rho)$. Equations (7.6)–(7.8) also describe helium-3, but with different material properties.

In accordance with chapter 3, we write the equations of state for $He^4$ and $He^3$ at 0.15 K

$$p = 7.57 \cdot (\rho - 93.24)^3 - 9.5 \cdot 10^5 [\text{Pa}] \;\left(\text{He}^4\right), \tag{7.9}$$

$$p = 13.7 \cdot (\rho - 51.55)^3 - 3 \cdot 10^5 [\text{Pa}] \;\left(\text{He}^3\right). \tag{7.10}$$

Along with equations (7.9)–(7.10), these allow us to describe the dynamics of liquid helium in each phase near a needle-like electrode, as we have done for water in chapter 5. The boundary conditions for equations (7.6) and (7.7) have a standard form: no-slip condition on the electrode surface, where the fluid velocity tends to zero, and continuity conditions of density and momentum fluxes on the boundaries of the computational domain.

Because a maximum electrostrictive tension occurs near the tip of the needle-like electrode in the dielectric fluid, as an example we will consider this electrode as a sphere with a radius equal to the radius of curvature of the electrode's tip.

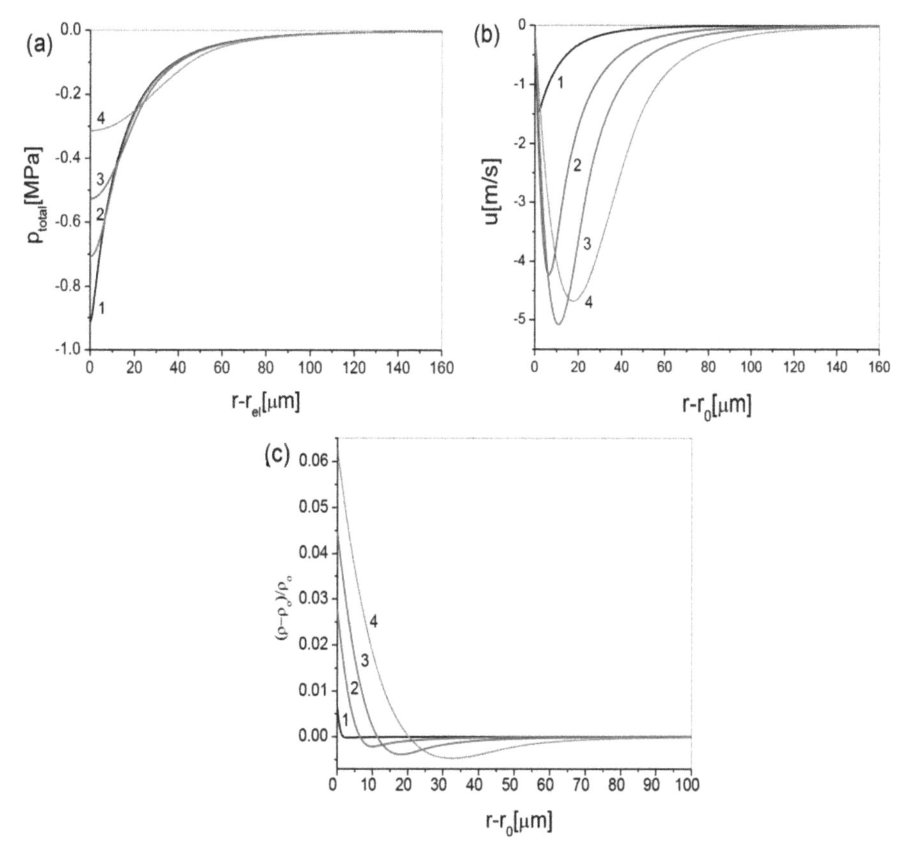


**Figure 7.1.** Results for helium-4. Radial distributions of total pressure $p_{\text{total}} = p + P_{\text{E}}$, fluid velocity, and fluid density for a spherical electrode with radius 50 μm and a maximum reached voltage of $U_0 = 100$ kV at time $t = \tau_0$. Curve 1 corresponds to $\tau_0 = 10$ ns, curve 2 to $\tau_0 = 50$ ns, curve 3 to $\tau_0 = 100$ ns, and curve 4 to $\tau_0 =$ 200 ns.

Figures 7.1 and 7.2 show the results of numerical calculations for helium-4 and helium-3 in the electrical field of a spherical electrode with radius $r_{\text{el}} = 50$ μm. The voltage on the electrode increases linearly with time and reaches a maximum of $U_0 = 100$ kV for helium-4 and $U_0 = 50$ kVfor helium-3 at time $\tau_0$. Accordingly, the electric field in the vicinity of the electrode is given by

$$E(r, t) = \frac{U_0 r_{\text{el}}}{r^2} \frac{t}{\tau_0}. \tag{7.11}$$

The electrostriction pressure $P_{\text{E}} = -\beta_{\text{E}} E^2$ practically coincides with curves 1 in figures 7.1(a) and 7.2(a). A decrease of the negative pressure with an increase of the voltage pulse duration (figures 7.1(a) and 7.2(a)) corresponds to the fluid flowing to the region of greater electric field. As a result, the fluid density (figures 7.1(c) and 7.2(c)) and the hydrostatic pressure $p$ increase. Figures 7.1(b) and 7.2(b) show that the maximum flow velocity first increases during the increase of pulse duration and then decreases, always remaining lower than the speed of sound.

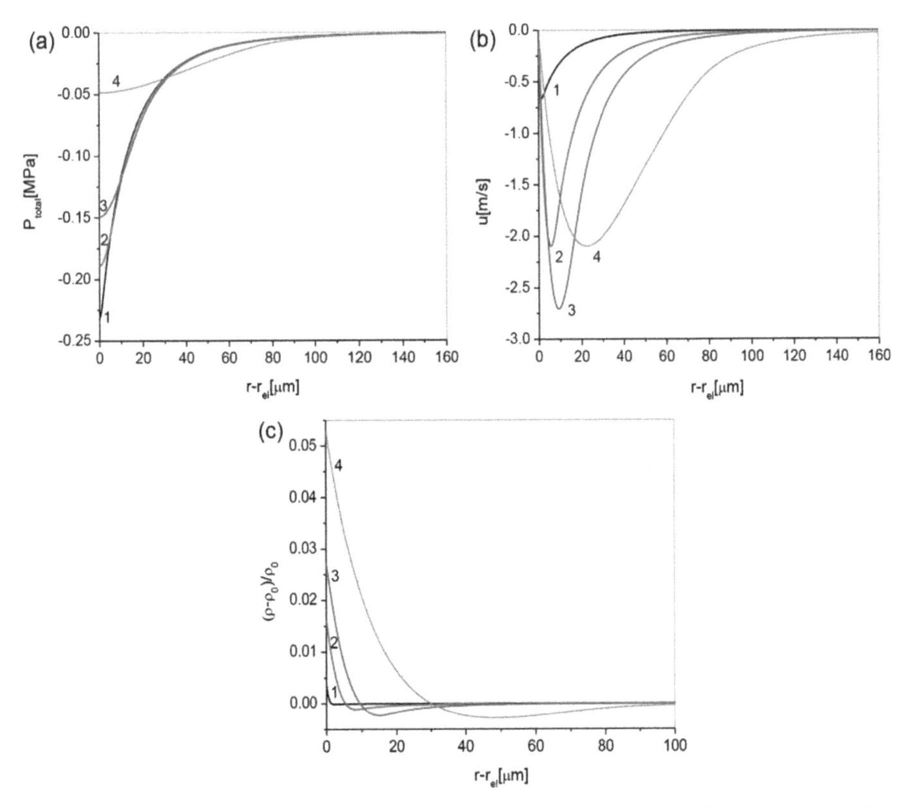


**Figure 7.2.** Results for helium-3. The same as figure 7.1, but for helium-3, with a maximum voltage of $U_0 = 50$kV, and curve 4 corresponds to $\tau_0 = 400$ ns.

Experimental data on cavitation occurrence in helium-4 and helium-3 at different temperatures are given in [11] (figure 7.3). The top and bottom curves in both graphs of figure 7.3 show the bounds of critical negative pressure $P_{-}(T)$ obtained by using different experimental techniques. For helium-4, the negative pressure at which cavitation occurs lies in the range of –0.75 to –1 MPa, and for helium-3 it is between –0.2 and –0.3 MPa.

Figures 7.1 and 7.2 show that the main conditions for the formation of cavitation are fulfilled at the voltage pulse duration of $\tau_0 < 10$ ns with an amplitude of 100 kV for helium-4 and 50 kV for helium-3 across a needle-like electrode with a curvature radius of 50 μm. At longer voltage pulse durations, cavitation should not occur, because the fluid inflow to the electrode compensates the electrostrictive tension.

Following equation (1.40), a good estimate of how quickly the voltage on the electrode should increase in order to neglect the liquid inflow to the high field region is $\tau_0 \ll \tau_P \approx 1.55 r_{el}/c_s$, where $\tau_0$ is the voltage rise duration and $\tau_P$ is the characteristic time for liquid inflow to the electrode. If $\tau_0 \ll \tau_P$, the hydrostatic pressure $p$ is less than $|P_E|$ and can be neglected. If, however, $\tau_0 \geqslant \tau_P$ ($p \approx |P_E|$), the total pressure, which is the sum of the hydrodynamic and electrostrictive pressures, may be insufficient for the formation of discontinuities in the liquid.

The speed of sound at a temperature of 0.15 K and atmospheric pressure in helium-4 is 238 m s$^{-1}$ and about 182 m s$^{-1}$ in helium-3 (figure 3.6). Hence, for an electrode of $r_{el} = 50$ μm, the voltage pulse duration $\tau_0$ at which cavitation may occur is less than 320 ns in helium-4 and $\tau_0 < 430$ ns in helium-3. Obviously, inequality $\tau_0 \ll \tau_p$ is satisfied with a large margin for pulse rise times of $\tau_0 < 10$ ns, which is necessary for cavitation inception at the considered conditions of figures 7.1 and 7.2. In accordance with equations (1.40) and (1.41), the maximum speed of fluid inflow to the electrode at $\tau_0 \approx \tau_p$ is of the order $|u_{max}| \approx 0.22|P_E|/\rho_0 c_s$. For the corresponding values of the electrostrictive negative pressure $P_E$, initial density $\rho_0$, and speed of sound $c_s$, this estimate yields a maximum flow velocity of $|u_{max}| = 6.4$ m s$^{-1}$ for helium-4 and $|u_{max}| = 3.7$ m s$^{-1}$ for helium-3, which is in good agreement with the numerical calculations.

At other temperatures, the constants in the equations of state (7.9) and (7.10) are different, but the results remain qualitatively the same.

## 7.2 Regimes of cavitation inception in liquid helium

The following formula defines the amount of cavitation bubbles emerging in a unit volume of liquid per unit volume per unit time (see (2.18)):

$$\frac{dn_b}{dt} = \Gamma = \frac{1}{V_{cr}} \cdot \frac{k_B T}{2\pi\hbar} \exp\left(-\frac{W_{cr}}{k_B T}\right), \tag{7.12}$$

where $W_{cr}$ is the minimum activation energy needed to create a growing bubble and $V_{cr} = \frac{4}{3}\pi R_{cr}^3$ is the initial volume of the bubble. $W_{cr}$ and $V_{cr}$ depend on the surface tension coefficient $\sigma_s$ and on the negative pressure $P_-$. Since the exponential term in (7.12) is inversely proportional to temperature, the nucleation rate should significantly reduce even with a slight decrease in ambient temperature. However, as experiments show, the negative pressure at which cavitation begins is almost independent of temperature at $T < 0.2$ K for both liquid $He^4$ and $He^3$ (figure 7.3) [1]. Moreover, in this temperature range and at about atmospheric pressure, helium-3 remains in a normal state, whereas helium-4 is superfluid.

From the classical point of view, such nucleation rate behavior in this temperature range is only possible in the case of barrier-free mechanisms of cavitation formation, i.e. when $W_{cr} \to 0$ at a certain value of the negative pressure $P_-$. In chapter 2, we considered a simple phenomenological model of allowing barrier-free formation of cavitation nanopores, justified by the fact that the empirical coefficient of surface tension $\sigma_s(R_{cr}) \to 0$ at $R_{cr} \to 0$, as described by equation (2.9) or (2.10).

However, it turns out that quantum nanopore nucleation explains the observed independence of critical negative pressure on temperature in liquid helium at $T \to 0$. As mentioned above, in the classical case, the fluctuation occurrence probability of the critical nanopore follows the Arrhenius law [15], i.e. $\Gamma \propto \exp(-\frac{W_{cr}}{k_B T})$. But in the quantum case, which is liquid helium at $T \to 0$, it becomes possible to overcome the activation barrier by tunneling. In this case, the tunneling probability determines the occurrence probability of a critical size pore: $\Gamma \propto \exp(-S)$, where $S$ is the

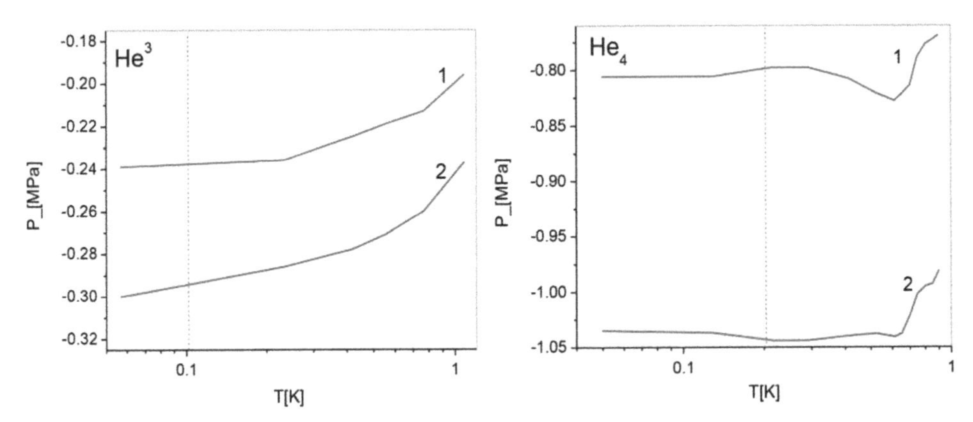


**Figure 7.3.** The upper (curves 1) and lower (curves 2) limits of the measured negative critical pressure at which cavitation occurs in $He^3$ and $He^4$ [1]. The vertical dashed lines separate the classical (right) and quantum (left) regimes of cavitation formation.

minimum of the imaginary-time action [7, 10, 16]. Many papers have studied and discussed the quantum nucleation of cavitation in the process of tunneling in detail (e.g. [7, 10, 18]).

At higher temperatures, the classical case takes over the nucleation process. So at some critical temperature $T*$, the probability of the classical thermal activation of cavitation (by the Arrhenius law) equals the quantum tunneling probability. Maris [10] and Guilleumas *et al* [18] obtained estimates of $T* \approx 0.1$ K for $He^3$ and $T* \approx 0.2$ K for $He^4$. The dashed vertical lines in figure 7.3 mark these values. The dependence of $T*$ on the magnitude of the critical negative pressure is almost independent for both liquids.

## 7.3 Possible limitations associated with the dielectric strength of liquid helium

In the first section of this chapter, we showed the fundamental possibility of cavitation initiation in liquid helium by electrostrictive tensions arising in pulsed inhomogeneous electric fields. This effect requires the field's amplitude to be $>10^9$ V m$^{-1}$, which is an order of magnitude greater than measured values for static or microsecond pulse breakdown fields in liquid helium [19–21]. Experiments with water and other polar and non-polar dielectric fluids in nanosecond and sub-nanosecond pulsed fields show a rise in the dielectric strength threshold by more than an order of magnitude (see chapter 9 and the references cited therein).

Considering that the required characteristic rise time of the voltage pulse is <10 ns (even better if the rise time <1 ns), one may hope that electrical breakdown does not have time to develop. In the recent theoretical work [22] a streamer breakdown in liquid helium was studied and the tendency of the breakdown field to increase as the duration of the voltage pulse decreases was shown. Unfortunately, we did not find any published data on the dielectric strength of liquid helium in the nanosecond range. Therefore, it is desirable to experimentally investigate the breakdown characteristics of liquid helium in the nanosecond and sub-nanosecond ranges in order to draw final conclusions about the possibility of cavitation production by a pulsed non-uniform electric field. Perhaps, such cavitation will promote the development of electric breakdown in liquid helium at nanosecond and sub-nanosecond voltage pulses, as happens in water, for example (see chapter 9).

## 7.4 Conclusions

1. The hydrodynamic equations describing the dynamics of liquid helium in a non-uniform nanosecond pulsed electric field coincide with the corresponding equations for water and other polar or non-polar dielectric liquids. In other words, liquid helium is a 'conventional fluid' regardless of whether it is in the normal or superfluid state.
2. If the total pressure (hydrostatic plus electrostrictive) is negative in the volume of liquid helium, and its absolute value exceeds a certain critical level, then cavitation arises in this volume, as happens in any other dielectric liquid.
3. Verification of the results in this chapter requires experimental measurements of the dielectric strength and electrostrictive phenomena in liquid helium in nanosecond and sub-nanosecond pulsed non-uniform electric fields.

## References


[1] Caupin E and Balibar S 2001 Cavitation pressure in liquid helium *Phys. Rev.* B **64** 064507

[2] Finch R D, Kagiwada R, Barmatz M and Rudnick I 1964 Cavitation in liquid helium *Phys. Rev* **134** A1425

[3] Jarman P D and Taylor K J 1970 The sonically induced cavitation of liquid helium *J. Low Temp. Phys.* **2** 389

[4] Murakami M and Harada K 2009 Thermo-fluid dynamic experiment of He II cavitating flow *Proc. 7th Int. Symp. on Cavitation (Ann Arbor, MI, August 17–22, 2009)* paper No. 65

[5] Blazkova M, Chagovets T V, Rotter M, Schmoranzer D and Skrbek L 2007 Cavitation in liquid helium due to a vibration quartz fork *Colloq. Fluid Dynamics 2007 (Prague, October 24–26)*

[6] Maris H and Balibar S 2000 Negative pressures and cavitation in liquid helium *Phys. Today* **53** 29

[7] Lifshitz I M and Yu Kagan 1972 Quantum kinetics of phase transitions at temperatures close to absolute zero *Sov. Phys.—JETP* **35** 206

[8] Casulleras J and Boronat J 2000 Progress in Monte Carlo calculations of Fermi systems: normal liquid $^3$He *Phys. Rev. Lett.* **84** 3

[9] Bauer G H, Ceperley D M and Goldenfeld N 2000 Path-integral Monte Carlo simulation of helium at negative pressures *Phys. Rev.* B **61** 9055

[10] Maris H J 1995 Theory of quantum nucleation of bubbles in liquid helium *J. Low Temp. Phys.* **98** 403

[11] Campbell C E, Folk R and Krotscheck E 1996 Critical behavior of liquid 4He at negative pressures *J. Low Temp. Phys.* **105** 13

[12] Shneider M N and Pekker M 2016 Rayleigh scattering on the cavitation region emerging in liquids *Opt. Lett.* **41** 1090

[13] Shneider M N and Pekker M 2015 Pre-breakdown processes in a dielectric fluid in inhomogeneous pulsed electric fields *J. Appl. Phys.* **117** 224902
[14] Pekker M and Shneider M N 2015 Pre-breakdown cavitation nanopores in the dielectric fluid in the inhomogeneous, pulsed electric fields *J. Phys. D: Appl. Phys.* **48** 424009
[15] Jackson H W 1982 Electrostriction in $^4$He *Phys. Rev.* B **25** 3127
[16] Jackson H W 1979 Foundations of a comprehensive theory of liquid He4 *Phys. Rev.* B **19** 2556
[17] Jackson H W 1978 Variational methods in the hydrodynamic theory of liquid $^4$He *Phys. Rev.* B **18** 6082
[18] Guilleumas M, Barranco M, Jezek D M, Lombard R J and Pi M 1996 Quantum cavitation in liquid helium *Phys. Rev.* B **54** 16135
[19] Blank C and Edwards M H 1960 Dielectric breakdown of liquid He *Phys. Rev.* **119** 50
[20] Gerhold J 1989 Breakdown phenomena in liquid helium *IEEE Trans. Electr. Insul.* **24** 155
[21] Suehiro J, Yamasaki K, Matsuo H and Hara M 1994 Pulsed electrical breakdown in liquid helium in the μs range *IEEE Trans. Diel. Electr. Insul.* **1** 405
[22] Belevtsev A A 2015 Electronic transport coefficients and electric breakdown in condensed helium *High Temp* **53** 779

**IOP** Publishing

Liquid Dielectrics in an Inhomogeneous Pulsed Electric Field

**M N Shneider and M Pekker**

# Chapter 8

## Optical diagnostics in dielectric liquids in inhomogeneous pulsed fields

*This chapter presents the shadowgraph method and its Schlieren modification for investigating the initial and developed stages of cavitation. It also shows that Rayleigh scattering off nanopores allows the detection of cavitation in the early stages of its inception with spatial and temporal resolutions not available with other optical detection methods.*

As shown in previous chapters, under the influence of ponderomotive forces, liquid tends to set in motion to compensate the electrostrictive stretching tension (negative pressure). This gives rise to density perturbations, which are very small, because of the 'incompressibility' of liquid dielectrics. Nevertheless, modern optical techniques allow registration of such perturbations in the fluid with a high spatial and temporal resolution. The role of optical methods in the development of gas and fluid mechanics is well known and is described in a number of review articles and books (e.g. [1–3]). First, there are the shadow and interference techniques. Since the 1960s laser-based shadow and interference methods have been actively developed. Lasers have significant advantages over conventional non-coherent light sources due to their coherence, unidirectionality, and monochromaticity. Lasers have not only significantly expanded the possibilities of standard optical techniques such as shadow and Schlieren methods, but also led to new effective methods: holography, speckle photography, and laser Doppler anemometry.

Most conventional optical diagnostic methods in hydrodynamics are based on the fact that even a small change in the density leads to a noticeable change in the refractive index. At the same time, the diversity of modern optical methods can be distinguished into three main groups: the shadow methods, Schlieren methods, and various realizations of the optical interferometry method.

doi:10.1088/978-0-7503-1245-5ch8  

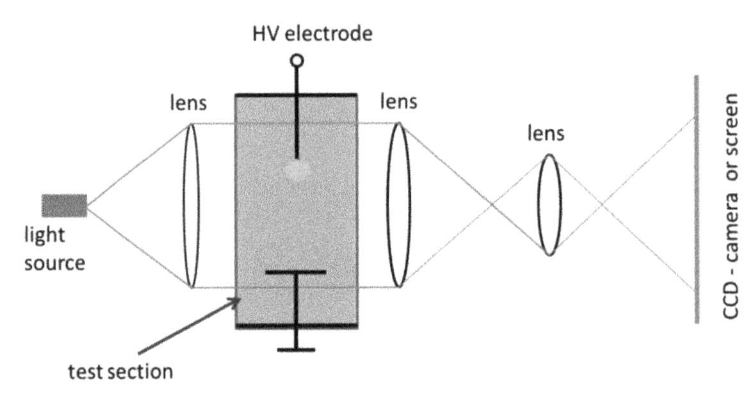


**Figure 8.1.** Schematic diagram of the shadowgraph technique.

## 8.1 Shadowgraph and Schlieren methods

The simplest, but still very effective, method is the shadowgraph [2–5]. Figure 8.1 shows the typical scheme of the shadow method. Shadowgraphs are created by passing a parallel beam of light through the test area in the liquid. Variations of the refractive index deflect the light rays and the degree of deflection depends on the local density perturbations. Deflected light rays produce bright and dark (shadows) areas in an image, called a shadowgraph, on a screen, photographic plate, or a sensitive digital charge-coupled device (CCD) camera.

Figure 8.2 shows examples of shadowgraphs of the rarefaction regions in the vicinity of a needle-like electrode in water and ethanol, when the voltage pulse amplitudes are below the breakdown threshold. The images show a late stage after the high-voltage pulse when the rarefaction wave propagates from the electrode. Recall that liquid flows to the electrode during the pulse, resulting in a growth of the hydrostatic pressure, which compensates the negative pressure caused by the electrostrictive forces. At the back front of the high-voltage pulse and after it, the liquid moves away from the electrode (see chapter 5). These moments are shown in figure 8.2.

Figure 8.3 [7] shows an example of shadow pictures of cavitation in water near the pin electrode at a pre-breakdown voltage pulse. It is worth noting that cavitation is emerging not only during the high-voltage pulse, but also considerably after the termination of the pulse, when the liquid flows away from the electrode creating a region of negative pressure as shown in the numerical simulation of figure 5.6.

Figure 8.4 [8] presents another example of shadowgraphs of nanosecond discharge in water.

The Schlieren method is very similar to the shadow method [4, 5]. It also uses a point light source at the focus of a lens, creating a parallel beam passing through the test area. A diaphragm (or knife-edge) is placed behind the test area to partially block the light beams that pass through this area (figure 8.5). Usually, the Schlieren technique is applied to visualize the density gradients in liquids and gases.

Interferometry relies on the fact that the local density changes not only lead to deviations of light rays, but also to a phase shift [1, 3, 9]. Usually, interference occurs

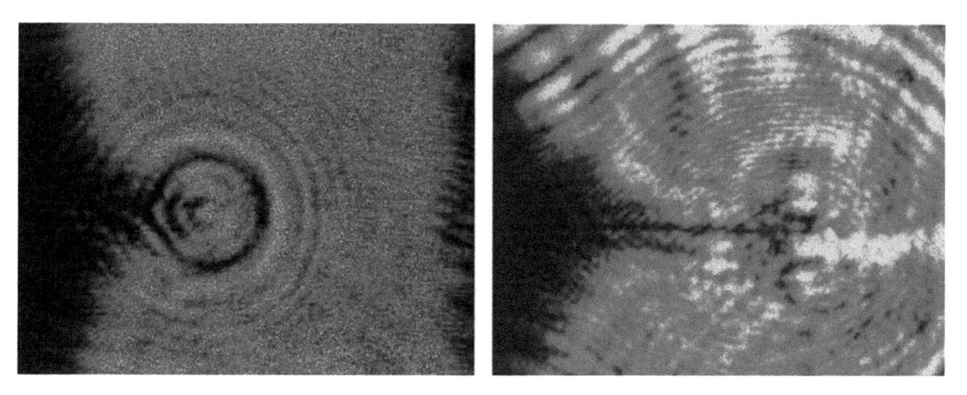

**Figure 8.2.** Shadowgraphs of a rarefaction wave without breakdown induced by electrostriction. The electrode tip radius is approximately 5 μm. The left panel shows water captured 200 ns after a high-voltage pulse of $U_0 = 9$ kV with a duration of 20 ns. The right panel shows ethanol 1000 ns after a pulse of $U_0 = 20$ kV and 60 ns. Reproduced with permission from [6].

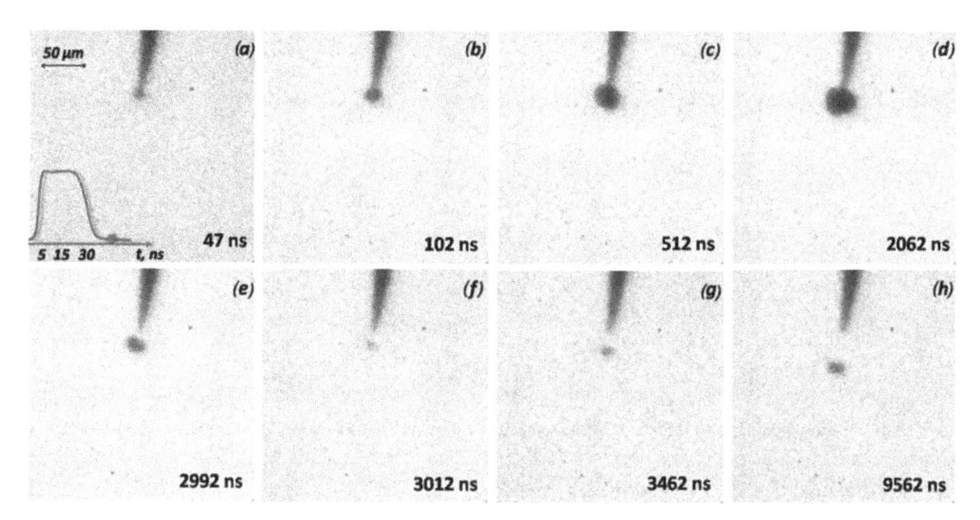


**Figure 8.3.** Cavitation formation in deionized water at a 4 kV high-voltage pulse with a rise time of 3.7 ns and FWHM ~30 ns. The nickel pin electrode tip has a radius of 1 ± 0.5 μm. The single-shot images are captured by an intensified CCD camera with a gate of 2 ns. The time stamps in the frames correspond to instants from the voltage rise on the electrode. Reproduced from [7].

at the superposition of two light beams formed by splitting the initial beam. One beam passes through the test section, while the other (reference) beam does not interact with the liquid. Then, the two beams converge and are projected onto a screen or CCD camera. The result is an interference picture that shows a phase shift caused by the density perturbations in the test area.

The above methods are generally applicable to studies of processes in homogeneous liquids or in multiphase liquids in which the bubbles formed are considerably larger than the laser wavelength and where there is a pronounced liquid–gas boundary, such as in boiling liquid or at developed cavitation. Methods related to

**Figure 8.4.** Shadowgraphs and corresponding plasma emission visualization of nanosecond discharge in deionized water under a 15 kV positive pulse with a rise time of 3.7 ns and FWHM ~30 ns. The nickel pin electrode tip radius is $1 \pm 0.5$ µm. Reproduced from [8].

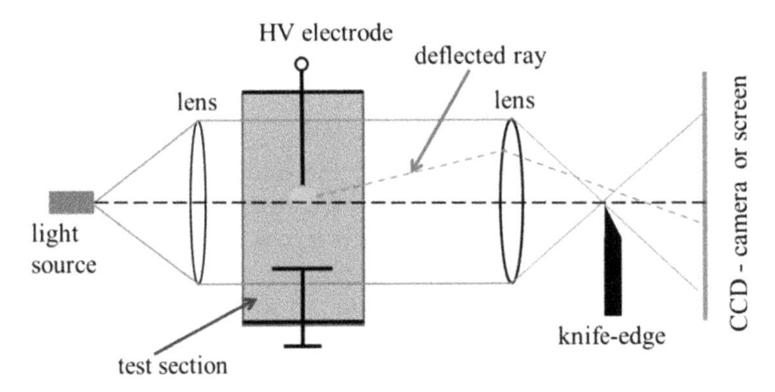


**Figure 8.5.** Schematic diagram of the Schlieren technique.

the scattering of light, on the other hand, are more efficient for observing transient processes in a liquid where a new phase emerges, such as cavitation in the area of negative pressure. Figure 8.6 depicts such an experiment, with cavitation emerging where converging sound waves form the negative pressure region at the stage of sound wave rarefaction in liquid helium-4 or -3, and a CCD camera or photomultiplier tube captures the laser radiation scattered by the cavitation bubbles [10, 11]. We will show below that this kind of scattering of emerging cavitation is Rayleigh scattering [12]. Scattering off larger cavitation bubbles that are comparable or exceeding the laser wavelength is Mie scattering [13].

Before turning to Rayleigh scattering, we will briefly describe the method of speckle photography. It is a modern laser technique that can be used to study electrostrictive processes in liquid dielectrics and is one of the promising methods for studying the initial state of emerging cavitation. It was developed in [14, 15] to study hydrodynamic processes in liquids and is reviewed along with its applications in [1, 16]. A speckle, or speckle pattern, is a random interference pattern that forms from the mutual interference of coherent waves with random phase shifts and/or a random set of intensities. A picture of such a pattern, projected onto a screen or recorded by a photosensitive digital camera, clearly shows bright spots (speckles), which are separated by dark areas. Figure 8.7 shows a typical experimental set-up of speckle photography. A planar laser beam illuminates the fluid in the region of interest. For minimal absorption, the laser wavelength should be in the visible light band, such as from a green argon laser, or a second harmonic of an Nd:YAG laser with a wavelength of 532 nm. The light scattered off a static fluid region, projected onto a screen or a digital camera, gives rise to a speckle pattern. Fluid motion or bubble formation (boiling or cavitation) cause changes in the speckle pattern. Considering that the typical speckle sizes with visible light are about 2–5 μm [1], this

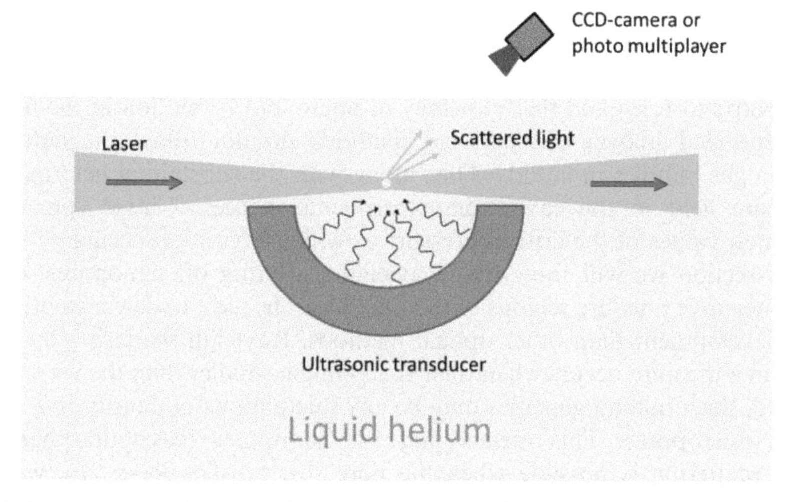


**Figure 8.6.** Sketch of an experiment in which cavitation forms at the focus of a hemispherical transducer due to the high amplitude pressure wing it creates. A laser beam passes through the acoustic focal region and a photomultiplier tube collects the light scattered off the bubbles.

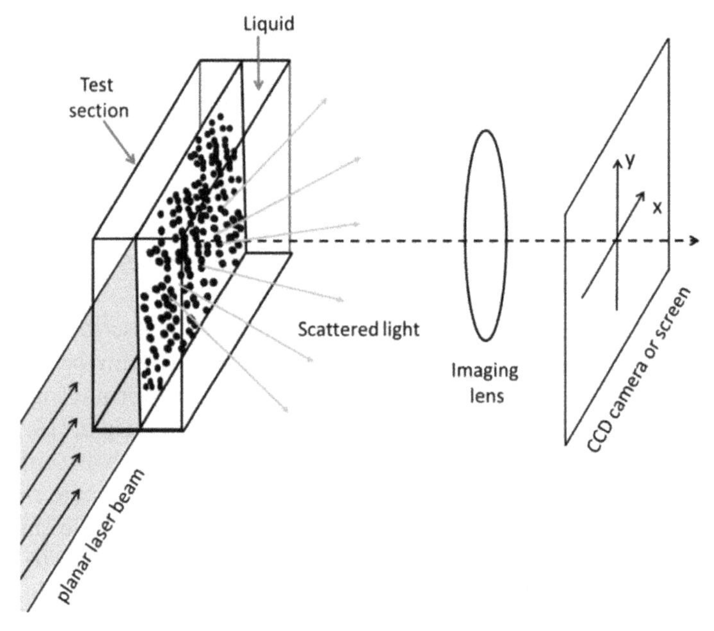


**Figure 8.7.** Experimental set-up of speckle photography, in which a thin sheet of light illuminates the liquid to produce the speckle pattern on a screen.

imaging technique may be good for the developed cavitation stage, but not quite suited to its very early stage of emerging nanovoids with sizes of the order of nanometers and tens of nanometers, i.e. much smaller than the laser wavelength.

## 8.2 Rayleigh scattering on the cavitation region emerging in liquids

As we mentioned in previous chapters, cavitation is a widespread natural phenomenon. One of the main problems of the study of cavitation in different environments is the experimental determination of the critical value of the negative pressure at which it starts to form and the frequency of microvoid formation in the liquid.

As mentioned above, conventional methods do not allow the detection of submicron gas bubbles in liquids. This, along with the substantial heterogeneity of the pressure field in the cavitation experiments, causes a large spread in the experimental values of the critical pressure at which cavitation occurs.

In this section we will show that Rayleigh scattering off nanopores, emerging from the negative pressure regions of the liquid, can be used to detect cavities earlier in their development than other optical methods. Rayleigh scattering by inhomogeneities in a medium occurs when their size is much smaller than the wavelength of light $\lambda$ [19]. Such inhomogeneities may be any fluctuations of density in a medium, including micropores. This means that for nanopores (cavitation nanovoids), Rayleigh scattering is possible when the pore size satisfies $R_b \ll \lambda/n$, where $n$ is the refractive index of the medium, and $\lambda$ is the wavelength of light in a vacuum. In this case, we can assume that a nanopore is located in the uniform electric field $E = E_0 \cdot e^{i\omega t}$, where $\omega = 2\pi c/\lambda$, and $c$ is the speed of light in a vacuum. In the frequency range of visible light, $n \approx 1.33$ for water. In such a field, a spherical cavity of radius $a$ behaves like an oscillating dipole with the dipole moment

$$\vec{p}_{\mathrm{b}} = \alpha \vec{E}_0 \cdot \mathrm{e}^{\mathrm{i}\omega t} \tag{8.1}$$

due to the periodic polarization of liquid on its borders. Here

$$\alpha = 4\pi\varepsilon_0 a^3 \frac{1 - n^2}{1 + 2n^2} \tag{8.2}$$

is the effective polarizability of the cavity in dielectric media [20], in which $\varepsilon \approx n^2$ is assumed for visible light in water.

To describe Rayleigh scattering in media, it is convenient to use the so-called scattering factor [20],

$$Y_{\mathrm{b}}(r,\, \theta) = \frac{I(r,\, \theta)}{I_{\mathrm{I}}}, \tag{8.3}$$

where $I_{\mathrm{I}} = \varepsilon_0 n c E_0^2/2$ is the intensity of the incident radiation, in which $E_0$ is the electric field amplitude, $I(r,\, \theta) = \frac{\omega^4 \alpha^2 \sin^2\theta}{16\pi^2 r^2 \varepsilon_0^2 c^4} I_{\mathrm{I}}$ is the intensity of the scattered radiation such that it makes the angle $\theta$ with respect to the induced dipole vector, and $r$ is the distance from the nanopore to the observation point [13, 22] (figure 8.8). The corresponding scattering factors in the direction determined by angle $\theta$ and integrating over the solid angle are

$$Y_{\mathrm{b}}(r,\, \theta) = \frac{I(r,\, \theta)}{I_{\mathrm{I}}} = \frac{\omega^4 \alpha^2 \sin^2\theta}{16\pi^2 r^2 \varepsilon_0^2 c^2} = \frac{\pi^2 \alpha^2 \sin^2\theta}{\lambda^4 \varepsilon_0^2 r^2} \tag{8.4}$$

$$Y_{\Omega,\mathrm{b}}(r) = \int Y_{\mathrm{b}}(r,\, \theta)\, \mathrm{d}\Omega = \frac{8\pi^3 \alpha^2}{3 r^2 \varepsilon_0^2 \lambda^4}. \tag{8.5}$$

If the cavitation nanopores are distributed randomly and the average distance between them is greater than the wavelength $l_{\mathrm{b}} > \lambda/n$, the scattered radiation is

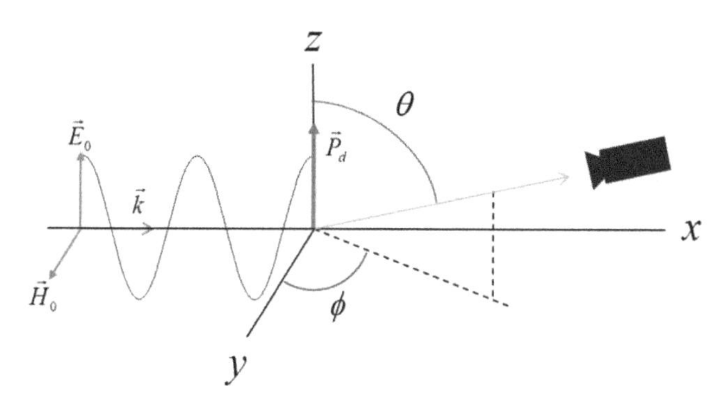


**Figure 8.8.** Scheme of Rayleigh scattering. $\vec{P}_{\mathrm{d}}$ is the induced dipole moment.

uncorrelated and the scattering factor is proportional to the number of pores $N_b$ in the irradiated scattering volume:

$$Y_{N,b}(r,\theta) = N_b \cdot Y_b(r,\theta) \tag{8.6}$$

$$Y_{N,\Omega,b}(r) = N_b \cdot Y_{\Omega,b}(r). \tag{8.7}$$

Scattering off cavitation micropores will be noticeable if it reaches or exceeds the level of Rayleigh scattering off the background liquid. For scattering by the background liquid molecules, when the characteristic size of the scattering region $L_s \gg \lambda/n$, the complete mutual interference quenching of radiation scattered by individual molecules holds and Rayleigh scattering is determined by thermal fluctuations [21, 23]. Note that in our recent paper [12], we erroneously compared the scattering of a plane-polarized electromagnetic wave off nanopores to the scattering of natural polarized light off the background fluctuations of the water. Below, we will correct this error.

The Rayleigh scattering factor for a small volume of liquid $V$, irradiated with a monochromatic planar electromagnetic wave, can be written in terms of $\theta$, the angle between the scattered and incident radiation (figure 8.8), and also in terms of scattering over the full solid angle:

$$Y_{\text{fluid}}(r,\theta) = \frac{\pi^2 V}{4\lambda^4 r^2}\left(\rho\frac{\partial\varepsilon}{\partial\rho}\right)_T^2 \beta_T k_B T \sin^2\theta, \tag{8.8}$$

$$Y_{\Omega,\text{fluid}}(r) = \int Y_{\text{fluid}}(\theta,\varphi,r)\,\mathrm{d}\Omega = \frac{2\pi^3 V}{3\lambda^4 r^2}\left(\rho\frac{\partial\varepsilon}{\partial\rho}\right)_T^2 \beta_T k_B T. \tag{8.9}$$

Here $\beta_T = -(\frac{1}{V}\frac{\partial V}{\partial p})_T = 4.8 \cdot 10^{-10}\ \text{m}^2\,\text{N}^{-1}$ is the coefficient of volume expansion of water, $k_B$ is the Boltzmann constant, and $(\rho\frac{\partial\varepsilon}{\partial\rho})_T \approx 1$.

The ratio of the Rayleigh scattering intensities off cavitation micropores in volume $V$ and off water of the same volume is independent of scattering angles and follows from (8.6)–(8.9):

$$\xi = \frac{Y_{N,b}(r,\theta)}{Y_{\text{fluid}}(r,\theta)} = \frac{Y_{N,\Omega,b}}{Y_{\Omega,\text{fluid}}} = \frac{64\pi^2}{\left(\rho\frac{\partial\varepsilon}{\partial\rho}\right)_T^2 \beta_T k_B T}\left(\frac{1-n^2}{1+2n^2}\right)^2 n_b a^6 \approx 9\cdot 10^{-6} n_b a^6. \tag{8.10}$$

Here, $n_b = N_b/V$ is the density of the nanopores, $a$ is the pore radius in nanometers, and the numerical coefficients correspond to the water temperature of 300 K. Since $l_b > \lambda/n$ holds for Rayleigh scattering off cavitation bubbles, considering that $n_b \sim (\lambda/n)^{-3}$, we obtain $\xi \approx 8 \cdot 10^{-5} a^6$ for green light ($\lambda = 532$ nm). That is, when the size of the nanopores is $a > 4.8$ nm in a volume illuminated by a laser, the Rayleigh scattering off the nanopores exceeds the scattering off the thermal fluctuations of water in the same volume.

Following the simple physical theory of nucleation considered in chapter 2, the rate of cavitation void formation in a unit volume per second is

$$\frac{dn_b}{dt} = \Gamma = \frac{3k_B T}{16\pi(k_\sigma \sigma_{0s})^3} \frac{|P_-|^3}{4\pi\hbar} \exp\left(-\frac{16\pi(k_\sigma \sigma_{0s})^3}{3k_B T P_-^2}\right), \tag{8.11}$$

where $P_-$ is the local instantaneous negative pressure in the liquid. The parameter $k_\sigma$ characterizes the dependence of the surface tension coefficient on the critical radius of the nanovoids. If, for instance, we assume the Tolman approximation for $k_\sigma$ (2.9) and that the critical pressure in water at which cavitation occurs is –30 MPa, then $k_\sigma \approx 0.25$ follows from our estimations in [24, 25].

The number of cavitation micropores grows exponentially so, after reaching a certain density, consideration of micropore formation has to take into account the feedback reaction on the pressure in the liquid [26] (see chapter 2).

We will assume, for example, that the pulsed field has a linear rising front $E(t) = E_0 t/\tau_0$ where $\tau_0 = 4$ ns, and $E_0 \approx 2.47 \cdot 10^8$ Vm$^{-1}$ corresponds to the local negative pressure $|P_-| \approx |P_E| = 33$ MPa. Where, as shown in chapter 6,

$$P_E = -\alpha_E \varepsilon_0 \varepsilon E_0^2 / 2 \tag{8.12}$$

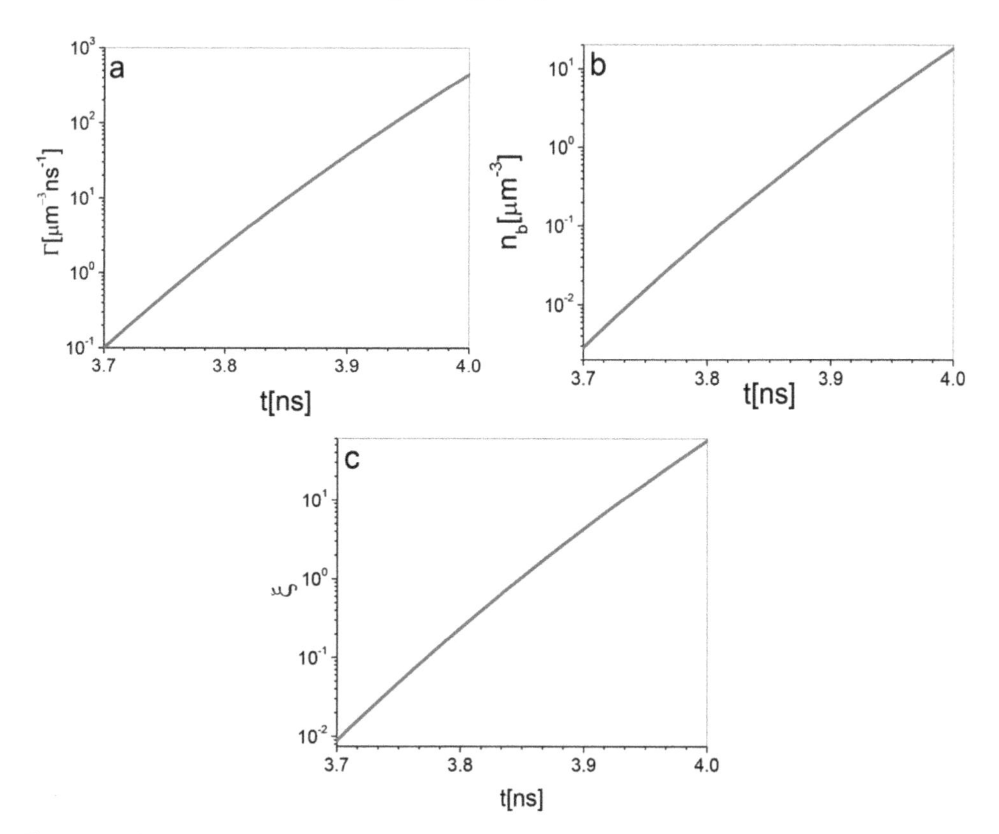


**Figure 8.9.** Time dependence of the rate of generation of cavitation voids, the number of pores, and parameter $\xi$ at the needle electrode, in which $E_0 \approx 2.47 \cdot 10^8$ Vm$^{-1}$ at $\tau_0 = 4$ ns.

Figure 8.9 shows the time dependence of the rate of generation of cavitation voids, the number of pores, and the parameter $\xi_{\theta=\pi/2}$ calculated by formulas (8.10) and (8.11). In these calculations we assumed that the nanopores are of equal size of 10 nm and $k_\sigma = 0.25$.

Note that for a time of the order of tens of nanoseconds, pores can grow to much larger sizes (the rate of the expansion of nanopores is about 100–300 m $s^{-1}$ [24, 25]), but we neglect this fact for simplicity. If, in the process of growth, the size of the pores reaches the order of the laser wavelength, then the scattering ceases to be isotropic Rayleigh and becomes anisotropic Mie scattering (e.g. [13, 19]), which we do not consider.

The combined use of Rayleigh scattering with the Schlieren method (which is a modification of the shadowgraph technique [3–5]), in conjunction with subtraction of the background laser scattering, allows the detection of nanopores of smaller size and lower concentration in the test volume. The Schlieren method was used in [27–30] to study the pre-breakdown stage in water and to detect cavitation development in the vicinity of a needle electrode. We considered this experimental scheme previously in chapter 5 (figure 5.8).

As shown in [27], at a voltage on the electrode of $U_0 > 10$ kV (with an electrode tip radius of curvature of $r_{el} \approx 35$ μm, and the corresponding negative pressure of $|P_E| > 36$ MPa (8.12)), there is a crescent-shaped area adjacent to the electrode that scatters the laser beam intensively. Figure 5.10 presents the corresponding Schlieren images (of size 340 × 230 $\mu m^2$) [27]. The 'dark' area appears in the vicinity of the electrode. The pressures shown in figure 8.10 correspond to the negative pressure near the electrode surface calculated by formula (8.12). It should be noted that the total pressure (sum of the electrostrictive negative pressure and hydrostatic

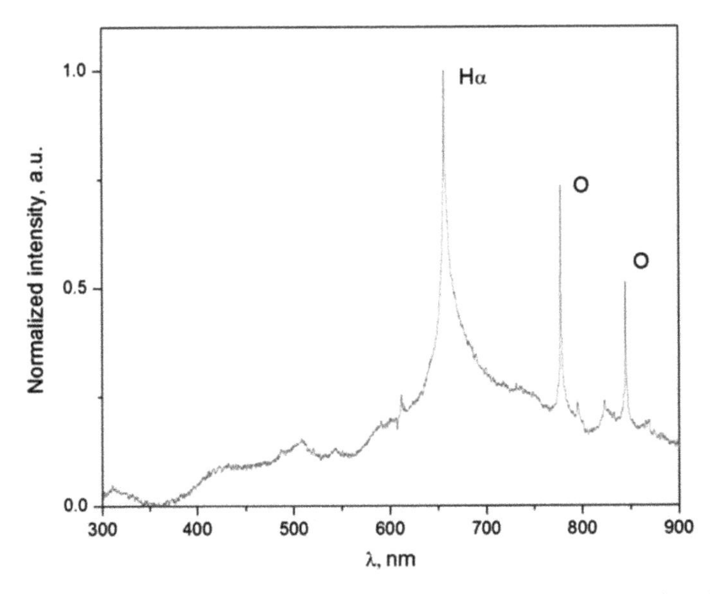


**Figure 8.10.** Spectrum of nanosecond discharge in liquid water. Reproduced from [28].

pressures) is lower than the absolute value of the negative pressure, because the fluid flows toward the electrode. In the bottom row of pictures in figure 8.10 (corresponding to the voltage amplitude of 9.5 kV), there are no noticeable dark areas adjacent to the electrode. This is apparently due to the fact that, in this case, the total pressure is less than critical for cavitation.

These 'dark' areas were observed up to $U_0$ = 19 kV and beyond that breakdown developed in water. The laser scattering areas in the vicinity of the electrode, observed in experiments [27], can be interpreted as Rayleigh scattering off the nanopores emerging from the cavitation. Certainty of this is only possible by gaining experimental confirmation of isotropy of the laser scattering in a plane perpendicular to the polarization direction.

Note that in the experiments [31], according to the results of model calculations, the absolute value of the negative pressure behind the shock wave initiated by the optical breakdown in water reached ~60 MPa and higher, which led to intensive cavitation development (figure 2.5 in chapter 2). As a result, the light scattered off the cavitation region looks similar to that observed in the vicinity of the electrode [27].

## 8.3 Optical emission spectroscopy of nanosecond and sub-nanosecond discharge in liquids

When breakdown develops in liquid, radiation emission in the visible and other spectral bands accompanies it. If this radiation is not absorbed in the liquid outside the discharge, it can be registered and provide information on the physical parameters of the liquid and the plasma generated in the breakdown zone. The spectral characteristics give information about the plasma, the electron and heavy particles temperatures, and the electron density. We will not dwell on the details of emission spectroscopy of breakdown in liquids. In conventional discharges in liquids with longer voltage pulses (a microsecond for example), plasma radiation is more likely to be generated in the gas bubbles. In this case, the spectral diagnostics of the discharge are not much different from the well-established diagnostics of weakly ionized gas discharge plasma (e.g. [32–34]).

The case of nanosecond breakdown is different, because bubbles and thus discharge in gas vapor just do not have time to form. Figure 8.10 shows a typical spectrum of a nanosecond breakdown in water [29]. The spectrum varies greatly at different moments of a high-voltage nanosecond pulse [35]. The discharge spectrum (figure 8.10) shows a strong broadening of hydrogen Balmer lines ($H_\alpha$ and $H_\beta$), almost continuum emission in the 300–900 nm region, and relatively weak broadened OI lines. It is possible to estimate the concentration of electrons in the plasma and the electron temperature from the analysis of Stark and collisional broadening, for example. Such analyses are presented in [8, 28] for discharge in water with a voltage pulse front duration of 3–5 ns. This analysis is performed similarly to the cases of conventional emission spectral diagnostics of gas discharge, or for breakdown in water at essentially longer voltage pulses, as in [36–38]. However, it should be noted that at the nanosecond breakdown of degassed water, the presence of intense spectral lines of neutral atoms (similar to those shown in figure 8.10) is not obvious and requires an explanation.

## References


[1] Lauterborn W and Vogel A 1984 Modern optical techniques in fluid mechanics *Ann. Rev. Fluid Mech.* **16** 223

[2] Merzkirch W 1987 *Flow Visualisation* 2nd edn (New York: Academic)

[3] Yang W J ed 1989 *Handbook of Flow Visualization* (New York: Hemisphere)

[4] Panigrahi P K and Muralidhar K 2012 *Schlieren and Shadowgraph Methods in Heat and Mass Transfer (Springer Briefs in Thermal Engineering and Applied Science)* (Berlin: Springer)

[5] Settles C S 2006 *Schlieren and Shadowgraph Techniques: Visualizing Phenomena in Transparent Media* 2nd edn (Berlin: Springer)

[6] Starikovskiy A 2013 Pulsed nanosecond discharge development in liquids with various dielectric permittivity constants *Plasma Sources Sci. Technol.* **22** 012001

[7] Marinov I L, Guaitella O, Rousseau A and Starikovskaia S M 2013 Cavitation in the vicinity of the high-voltage electrode as a key step of nanosecond breakdown in liquids *Plasma Sources Sci. Technol.* **22** 042001

[8] Marinov I, Starikovskaia S and Rousseau A 2014 Dynamics of plasma evolution in a nanosecond underwater discharge *J. Phys. D: Appl. Phys.* **47** 224017

[9] Verma S, Joshi Y M and Muralidhar K 2012 Optical interferometers: principles and applications in transport phenomena *Interferometry Principles and Applications* ed M E Russo (New York: Nova Science)

[10] Maris H and Balibar S 2000 Negative pressures and cavitation in liquid helium *Phys. Today* **53** 29

[11] Caupin E and Balibar S 2001 Cavitation pressure in liquid helium *Phys. Rev.* B **64** 064507

[12] Shneider M N and Pekker M 2016 Rayleigh scattering on the cavitation region emerging in liquids *Opt. Lett.* **41** 1090

[13] Born M and Wolf E 1980 *Principles of Optics* 6th edn (Oxford: Pergamon)

[14] Dudderar T D and Simpkins P G 1977 Laser speckle photography in a liquid medium *Nature* **270** 45

[15] Barker D B and Fourney M E 1977 Measuring fluid velocities with speckle patterns *Opt. Lett.* **4** 135

[16] Fomin N A 1998 *Speckle Photography for Fluid Mechanics Measurements (Experimental Fluid Mechanics)* (Berlin: Springer)

[17] Herbert E, Balibar S and Caupin F 2006 Cavitation pressure in water *Phys. Rev.* E **74** 041603

[18] Vinogradov V E 2009 Depression of the cavitation centers in water under pulsed tension conditions *Tech. Phys. Lett.* **35** 54–6

[19] Bohren C F and Huffman D R 1983 *Absorption and Scattering of Light by Small Particles* (New York: Wiley)

[20] Tamm I E 2003 *Fundamentals of the Theory of Electricity* (Moscow: Fizmatlit)

[21] Fabelinskii I L 1968 *Molecular Scattering of Light* (New York: Plenum)

[22] Miles R B, Lempert W R and Forkey J N 2001 Laser Rayleigh scattering *Meas. Sci. Technol.* **12** R33–R51

[23] Landau L D and Lifshitz E M 1960 *Electrodynamics of Continuous Media* (*A Course of Theoretical Physics* vol 8) (Oxford: Pergamon)

[24] Shneider M N and Pekker M 2015 Pre-breakdown processes in a dielectric fluid in inhomogeneous pulsed electric fields *J. Appl. Phys.* **117** 224902
[25] Pekker M and Shneider M N 2015 Pre-breakdown cavitation nanopores in the dielectric fluid in the inhomogeneous, pulsed electric fields *J. Phys. D: Appl. Phys.* **48** 424009
[26] Rosenberg L D 1968 The cavitation region *Powerful Ultrasound Waves* (Moscow: Nauka) chapter 6 (in Russian)
[27] Pekker M, Seepersad Y, Shneider M N, Fridman A and Dobrynin D 2014 Initiation stage of nanosecond breakdown in liquid *J. Phys. D: Appl. Phys.* **47** 025502
[28] Dobrynin D, Seepersad Y, Pekker M, Shneider M N, Friedman G and Fridman A 2013 Non-equilibrium nanosecond-pulsed plasma generation in the liquid phase (water, PDMS) without bubbles: fast imaging, spectroscopy and leader-type model *J. Phys. D: Appl. Phys.* **46** 105201
[29] Seepersad Y, Pekker M, Shneider M N, Dobrynin D and Fridman Al 2013 On the electrostrictive mechanism of nanosecond-pulsed breakdown in liquid phase *J. Phys. D: Appl. Phys.* **46** 162001
[30] Seepersad Y, Fridman A and Dobrynin D 2015 Anode initiated impulse breakdown in water: the dependence on pulse rise time for nanosecond and sub-nanosecond pulses and initiation mechanism based on electrostriction *J. Phys. D: Appl. Phys.* **48** 424012
[31] Ando K, Liu A-Q and Ohl C-D 2012 Homogeneous nucleation in water in microfluidic channels *Phys. Rev. Lett.* **109** 044501
[32] Fantz U 2006 Basics of plasma spectroscopy *Plasma Sources Sci. Technol.* **15** S137
[33] Lochte-Holtgreven W ed 1968 *Plasma Diagnostics* (Amsterdam: Wiley Interscience)
[34] Ochkin V N 2009 *Spectroscopy of Low Temperature Plasma* (Weinheim: Wiley)
[35] Fridman A, Dobrynin D, Friedman G, Fridman G, Cho Y, Pekker M and Shneider M 2013 Fundamental physics and engineering of nanosecond-pulsed nonequilibrium microplasma in liquid phase without bubbles *Final Report* AFRL-OSR-VA-TR-2013-0056
[36] Bruggeman P and Leys C 2009 Non-thermal plasmas in and in contact with liquids *J. Phys. D: Appl. Phys.* **42** 053001
[37] Bruggeman P, Schram D, Gonzalez M A, Rego R, Kong M G and Leys C 2009 Characterization of a direct dc-excited discharge in water by optical emission spectroscopy *Plasma Sources Sci. Technol.* **18** 025017
[38] Gucker S N, Foster J E and Garcia M C 2015 An investigation of an underwater steam plasma discharge as alternative to air plasmas for water purification *Plasma Sources Sci. Technol.* **24** 055005

**IOP** Publishing

Liquid Dielectrics in an Inhomogeneous Pulsed Electric Field

**M N Shneider and M Pekker**

# Chapter 9

## Breakdown in liquids in pulsed electric fields

*Experimental data of microsecond and nanosecond discharges in liquids are briefly reviewed. It is shown that the vapor bubble model of breakdown does not work for nanosecond high-voltage pulses. The cavitation model of sub-nanosecond and nanosecond breakdown initiation in liquid dielectrics is discussed.*

In the twentieth century, two fundamentally different approaches to the phenomenon of electric discharge in liquids were formed. According to the first one, an electrical discharge in a fluid is a discharge in gas cavities that are either already present in the liquid and on the electrodes, or are formed therein due to Joule heating, electrolysis, and boiling. This kind of breakdown and discharge is called a 'bubble' discharge. Ushakov, Klimkin, and Korobeynikov describe it in detail in the excellent monograph [1] where, together with an overview of the experimental data, they present a theory for the behavior of gas bubbles in strong electric fields. The drawback of the bubble model is the lack of self-consistent relations between the processes of ionization in the bubbles and the generation mechanism of the bubbles themselves in the pre-discharge and discharge stages of the breakdown. The bubble model mainly provides a qualitative picture of the discharge in a liquid.

In the second approach, the discharge in the liquid is considered to be a result of the avalanche multiplication of free charges in the liquid itself. In other words, it assumes that in strong electric fields the electrons in a liquid can acquire sufficient energy to ionize atoms and molecules. In essence, this approach is an extension of a well-developed theory of gas discharge in the liquid phase of matter. This approach is called 'an ionization' model. According to supporters of the ionization discharge mechanism, all cross-sections of the elementary processes (ionization, charge transfer, and elastic and inelastic electron scattering) in the fluid differ from those measured in the gas. Hence, one can develop a theory that describes the discharge in liquid by modifying the relevant cross sections (e.g. [2, 3]).

One of the important differences in the development of classical breakdown in liquid from the breakdown in gas is the significant time it takes vapor cavities to

doi:10.1088/978-0-7503-1245-5ch9

form that are large enough for electrons to gain sufficient energy for ionization multiplication. To model this fact, Ushakov introduced 'the discharge ignition delay time' $t_{\text{delay}}$ in [4] as the time period from the beginning of the voltage rise until the first registered discontinuities form in liquid near the high-voltage electrode. This time period is related to the emergence of vapor bubbles, the change in their shape, and the beginning of the discharge process.

The emergence of new high-power pulsed high-voltage sources [5, 6] has led to opportunities for studying the breakdown in liquids in the nanosecond and sub-nanosecond regimes.

The first experiments [7] showed surprising results: the electric field at which breakdown occurred was an order of magnitude lower than expected from estimates for dense liquids. We will show below that vapor bubbles cannot form at times of a few nanoseconds (section 9.3).

In [8, 9], we proposed the concept of a bubble-free cavitation breakdown in dielectric liquids at nanosecond and sub-nanosecond high-voltage pulses applied to a needle-like electrode. According to this approach the development of cavitation breaks, in which the electrons can gain enough energy to ionize water molecules, makes electric breakdown possible. We discussed the conditions for cavitation nanovoid occurrence in pulsed inhomogeneous electric fields in previous chapters.

Below, without going into detail, we will give an overview of the experimental data and describe the problems facing the ionization and bubble models of discharge development in liquids. We will also briefly discuss the perspectives of a cavitation model for breakdown in liquids at a nanosecond high-voltage pulse applied to a pin electrode.

## 9.1 A brief overview of the experimental data

### 9.1.1 Microsecond breakdown in liquids

Figure 9.1 shows an example of a discharge in water ($E$ = 80 MV m$^{-1}$), which starts with a bubble at an anode [1]. The delay time is $t_{\text{delay}}$ = 0.5 μs. A bush-like formation

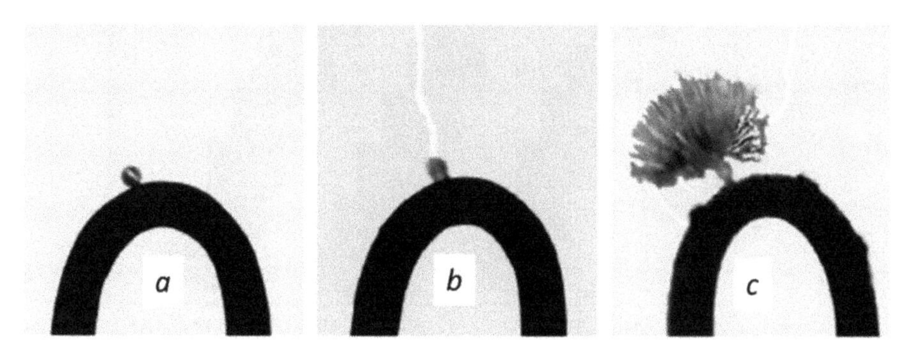


**Figure 9.1.** Breakdown development from an anode with an initial bubble. Panel (a) shows the initial bubble, (b) the discharge at the moment $t$ = 1.2 μs after the voltage pulse turns on, and (c) the fan (bush) of streamers that develop from the tip of the bubble at $t$ = 1.3 μs. The delay time of the discharge ignition is $t_{\text{delay}}$ = 0.5 μs. Reproduced with permission from [1]. Copyright 2007 Springer.

of discharge channels starts the tip of the bubble. The channels form a hemisphere of about 600 μm radius, the distance between their tips is 40–60 μm, and the total number of channels is apparently greater than a hundred. The propagation velocities of the channels are essentially subsonic, of order 100–400 m s$^{-1}$. In the absence of an initial bubble at the anode, the delay is $t_{delay} = 4$ μs [1].

Regarding the discharge from the cathode in the presence of bubbles, the delay time is $t_{delay} \approx 22$ μs, which is more than 40 times greater than the delay time in the anode discharge. Moreover, without bubbles on the cathode surface, discharge does not develop at all.

The influence of external pressure on breakdown in liquid was experimentally studied in [10, 11], and revealed that the dielectric strength substantially increased at elevated external pressures. These results confirm the important role of the formation of gas bubbles at the initial stages of discharge development.

We will show below that the observed breakdown time delays in microsecond voltage pulses are associated with the formation of vapor bubbles at the electrode due to Joule heating.

### 9.1.2 Nanosecond and sub-nanosecond breakdown in liquids

We present the basic experimental results from [7] without going into the details of the set-up. We describe these experiments schematically in chapter 5 (figure 5.8). Figures 9.2 and 9.3 show the dynamics of the light emission from the discharge area in distilled water at nanosecond and sub-nanosecond voltage pulses and of the high-voltage potential on the needle electrode, the tip of which has a radius of curvature of $r_{el} = 50$ μm, and the distance between the plate and the electrode is $d = 4$ mm.

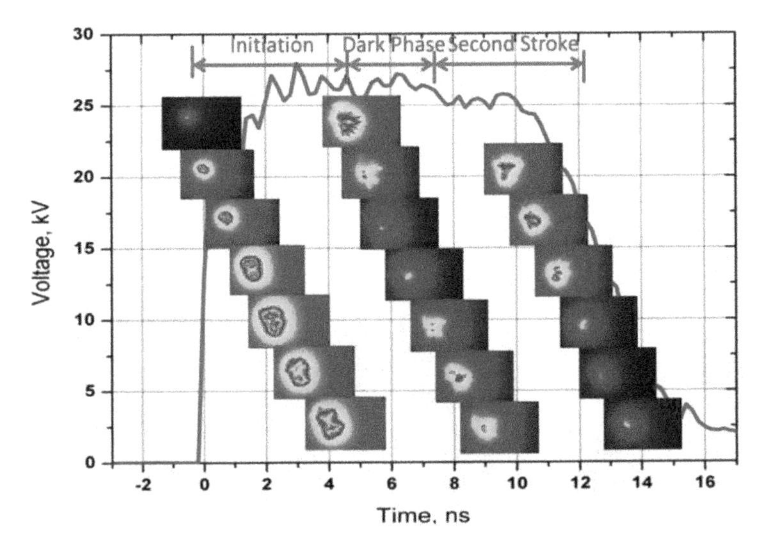


**Figure 9.2.** Dynamics of the discharge emission and high-voltage potential on the electrode in distilled water with a voltage pulse of $U_0 = 27$ kV. The camera gate is 1 ns and the spectral response is 250–750 nm. Reproduced from [7].

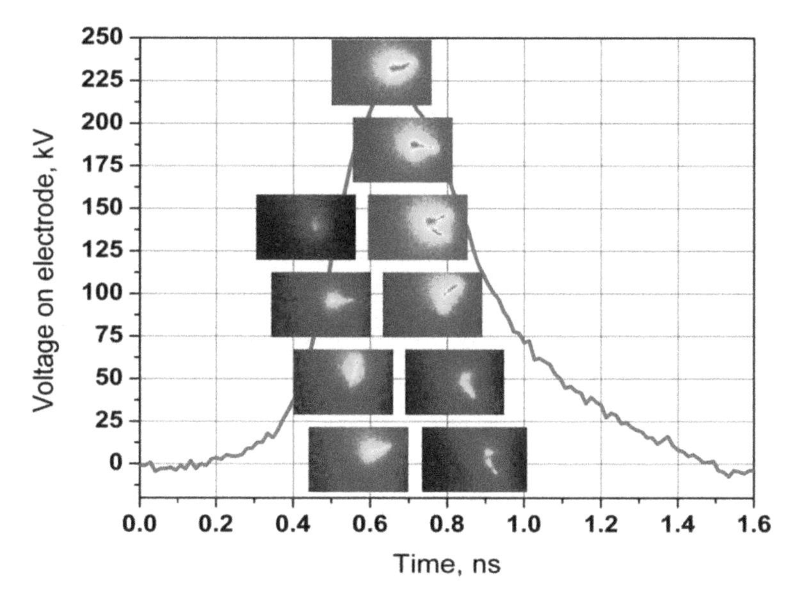


**Figure 9.3.** The same as figure 9.2, but with $U_0$ = 220 kV and $dU/dt$ = 1.46 MV ns$^{-1}$ [7]. The camera gate is 500 ps and the time shift between frames is 50 ps. Reproduced from [7].

The nanosecond discharge is ignited at 15 kV and develops (first stroke) at the front edge, then fades (dark phase), and is ignited again (second stroke) on the trailing edge of the voltage pulse (figure 9.2). At the sub-nanosecond voltage pulse (figure 9.3), the discharge is ignited at 100 kV. In this case, there is no dark phase.

Later, similar results were obtained in [12–18] for other electrode tip radii of curvature and nanosecond voltage pulse parameters.

Figure 9.4 shows the high-voltage pulses on the needle-like electrode in the experiments of [14] with a tip curvature radius of $r_{el}$ = 35 μm, and the distance between the plate and the electrode is $d$ = 1.5 mm. Figures 9.5(a)–(c) show the corresponding dynamics of the light emission from a discharge in water for a pulse of positive polarity with amplitude 23.1 kV [14]. Under these conditions, as shown in [17], when the voltage amplitude on the electrode is less than 20 kV, breakdown in water is not observed.

At the negative polarity on the high-voltage electrode, a nanosecond breakdown also takes place, but its development is significantly suppressed. Figure 9.6 shows the dynamics of the light emission from a discharge in water at the same conditions as in figure 9.5(a), but in the case of a negative voltage pulse on the needle electrode with the minimal value of –23.1 kV [14].

## 9.2 Problems with the ionization model of breakdown development in liquids

We estimate the critical field under which electrical breakdown becomes possible by considering the liquid as a super-dense gas. In a gas with an applied electric field, electrons can acquire energy during the mean free path and lose it in collisions with

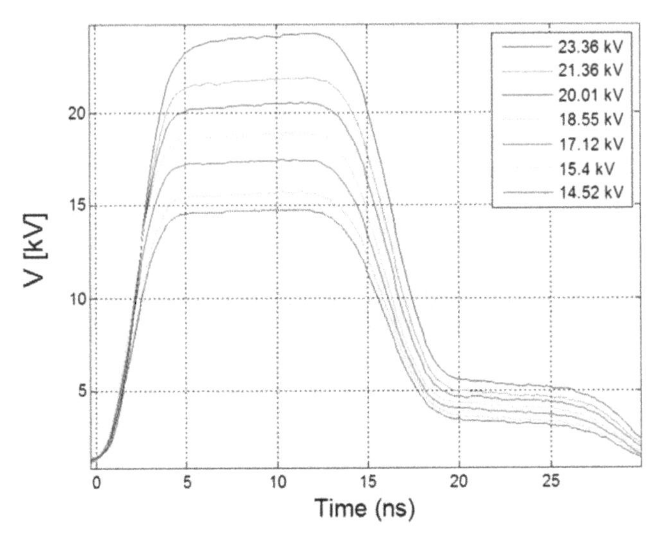


**Figure 9.4.** High-voltage pulses applied to the needle-like electrode. Reproduced from [14].

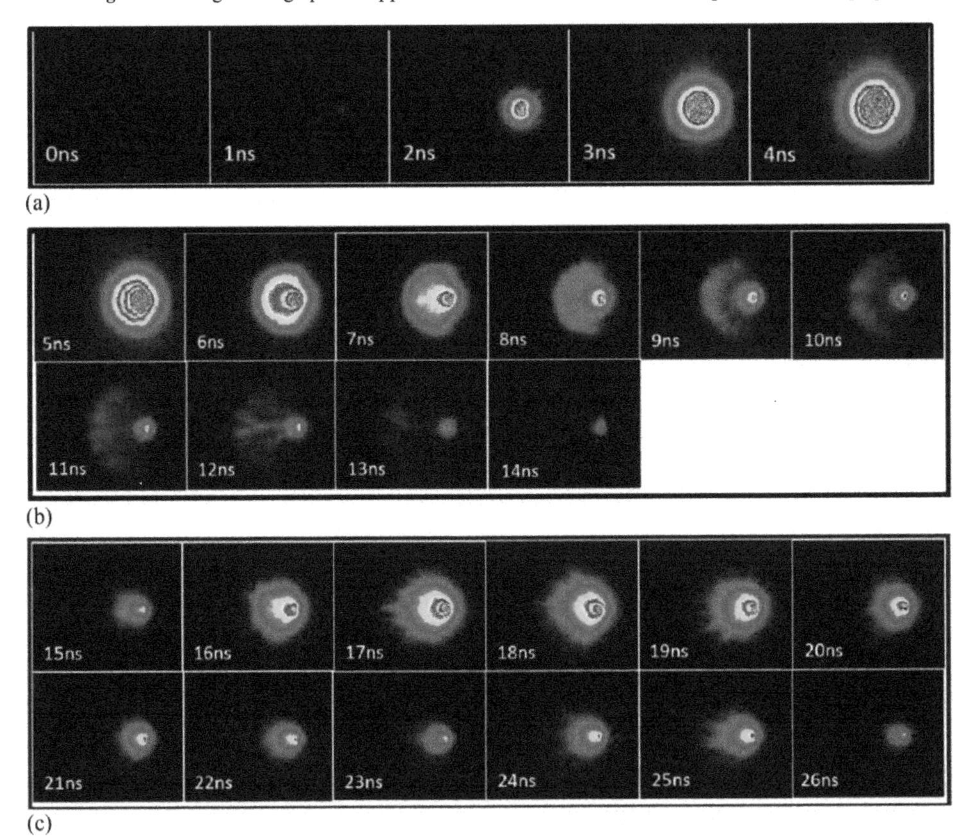


**Figure 9.5.** (a) Progression of the initial stage of discharge in water for the +23.1 kV voltage pulse from figure 9.4. These first 4 ns is the effective time over which the rising edge of the voltage pulse appears on the electrodes. (b) Continued progression of the discharge from (a) over the 10 ns during which the voltage is constant across the electrode gap. (c) Continued progression of the discharge from (a) and (b) over the falling edge of the applied pulse (15–20 ns) and beyond. Reproduced from [14].

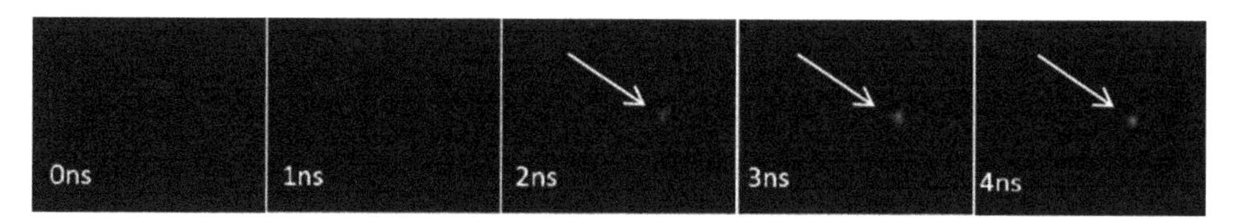


**Figure 9.6.** Development of the initial stage of discharge in water for the negative voltage pulse with the minimal value –23.1 kV. Reproduced from [14].

neutral molecules and charged particles. The breakdown conditions for discharge in gas follow the known criterion of linear dependence on the parameter $(E/N)_{br}$, where $N$ is the density of the gas [19]. Because the breakdown electric field in air at normal pressure and temperature is $E_{air} \approx 3 \cdot 10^6$ V m$^{-1}$ and the density of water is a thousand times greater than air density, one would expect-the breakdown field in water to be approximately three orders of magnitude greater than in air: $E_{water} \approx 3 \cdot 10^9$ V m$^{-1}$.

For practical estimations of the breakdown parameters in liquid dielectrics, Martin proposed a simple empirical formula for the breakdown electric field in terms of the voltage pulse duration, on the basis of experimental data for various liquids and a planar geometry of electrodes [20],

$$E_{br} = \frac{A_m}{\tau_0^{1/3} S^{1/10}}, \tag{9.1}$$

where $A_m$ is a fluid-dependent constant (in MV cm$^{-1}$), $\tau_0$ is the pulse duration (in μs), and $S$ is the area of the electrodes (in cm$^2$). For hexane and castor oil $A_m = 0.7$, for glycerol $A_m = 0.6$, for ethanol and transformer oil $A_m = 0.5$, and for water $A_m = 0.6$ in the cases of breakdown from the cathode and $A_m = 0.3$ in the anode case. Formula (9.1) assumes that the size of each electrode plate is much larger than the inter-electrode gap. Figure 9.7 shows the dependence of $E_{br}$ on the pulse duration, computed by (9.1) at $S = 1$ cm$^2$, for water, glycerol, and transformer oil.

Figure 9.7 illustrates that, for example, at a microsecond voltage pulse, the breakdown field in water is approximately three orders lower than the value that follows from the analogy with the breakdown in gas and its law of inverse proportionality to density.

## 9.3 Problems of the bubble breakdown model

In chapter 2, we showed that bubbles of radii smaller than the micron size cannot exist stably in liquid, due to surface tension at the liquid–bubble interface (Laplace pressure). Therefore, a bubble breakdown mechanism cannot cause the breakdown in degassed liquids at high-voltage pulse durations so short that the released heat is insufficient for bubble generation.

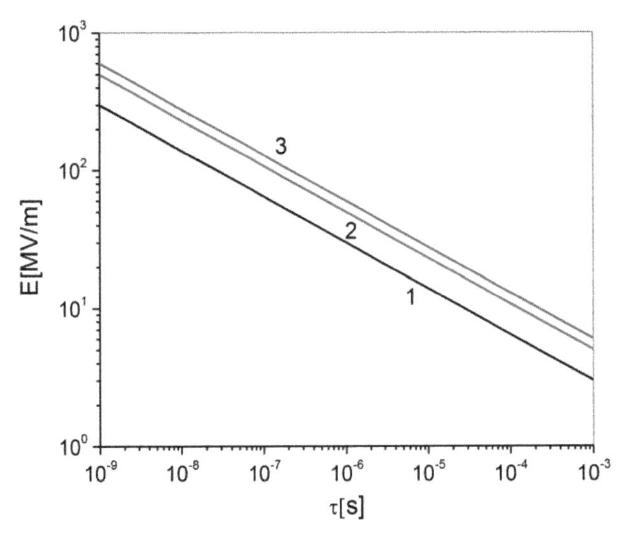


**Figure 9.7.** Dependence of the breakdown electric field on the voltage pulse duration across a planar capacitor with area $S = 1\ \text{cm}^2$, computed by Martin's formula (9.1). Curve 1 corresponds to water (the discharge begins from the anode), curve 2 to glycerol, and curve 3 to transformer oil.

We will now show that the nanosecond and sub-nanosecond electric breakdown observed in [7, 12–18] cannot be explained by the bubble model.

The amount of heat $W_T$ released per unit volume of liquid by Joule heating is

$$W_T(\vec{r}) = \int_0^{\tau_0} \left(\vec{j} \cdot \vec{E}\right) \mathrm{d}t = \Sigma \int_0^{\tau_0} E(\vec{r},\, t)^2 \mathrm{d}t, \tag{9.2}$$

where $\Sigma$ is the conductivity, $E(\vec{r},\, t)$ is the instantaneous electric field at point $\vec{r}$, and $\tau_0$ is the voltage pulse duration.

The field in the gap between a parabolic tip with curvature radius $r_{\text{el}}$ at potential $U$ and an ideally conducting plane electrode at zero potential a distance $d$ away is

$$E = \frac{2U}{(r_{\text{el}} + 2z)\ln\left(2d/r_{\text{el}} + 1\right)}, \tag{9.3}$$

in which $z$ is the distance from the tip to the plane electrode along the axis perpendicular to the plane [19].

In the vicinity of the electrode tip, where $z \ll r_{\text{el}}$,

$$E = \frac{2U}{r_{\text{el}} \ln\left(2d/r_{\text{el}}\right)}. \tag{9.4}$$

Here, we neglected the 1 inside the logarithm in (9.3), because $d/r_{\text{el}} \gg 1$.

Equation (9.2) is valid when $d \ll l_{\text{el}}$, where $l_{\text{el}}$ is the length of the needle-shaped electrode.

If the electrode is a perfectly conducting long cylinder with a semispherical tip, the electric field in the vicinity of the tip [21] is

$$E = \frac{U \cdot r_{\rm el}}{2r^2}. \tag{9.5}$$

In the gap between a sphere and a remote plane ($r_{\rm el}/d \ll 1$), the field is

$$E = \frac{U \cdot r_{\rm el}}{r^2}. \tag{9.6}$$

For an electrode of the ellipsoidal shape with semi-axes $r_{\rm el}$ and $a_z$, the corresponding expressions for the electric field are given in chapter 5.

Assuming that the voltage at the time interval $\tau_0$ varies linearly, equation (9.2) becomes the following at $r \approx r_{\rm el}$:

$$W_T = \frac{1}{3}\Sigma E_0^2(r)\tau_0, \tag{9.7}$$

in which $E_0$ is the maximum electric field at point $r$. For ultra-purified distilled water $\Sigma_{\rm dist} = 5 \cdot 10^{-6}\ {\rm S\ m^{-1}}$ and for regular tap water $\Sigma_{\rm tap} = 0.005–0.05\ {\rm S\ m^{-1}}$ [22].

Table 9.1 presents the experimental conditions under which the nanosecond discharges in water were observed in [7, 12, 14]. The electric field near the electrode was estimated by formula (9.3) and $P_- \approx P_{\rm E} = -\alpha_{\rm E}\varepsilon_0\varepsilon E^2/2$ was used for negative pressure (see chapter 5).

Substituting $E_0 \approx 0.3 \cdot 10^9\ {\rm V\ m^{-1}}$ and $\tau_0 = 2 \cdot 10^{-9}$ s from the above mentioned experiments into (9.7), we obtain the following for tap water and purified water, respectively:

$$W_{T,\rm distil} \approx 3 \cdot 10^2\ {\rm J\ m^{-3}}, \qquad W_{T,\rm tap} \approx 3 \cdot 10^5 - 3 \cdot 10^6\ {\rm J\ m^{-3}}. \tag{9.8}$$

On the other hand, the energy required to evaporate a cubic meter of water is $W_{\rm evap} = 2.2 \cdot 10^9\ {\rm J\ m^{-3}}$ and for electrolysis ($2H_2O \rightarrow 2H_2 + O_2$) $W_{\rm electr} = 1.4 \cdot 10^9\ {\rm J\ m^{-3}}$ [22]. That is, at a nanosecond voltage pulse, the energy released by Joule heating is so small that a vapor bubble cannot form.

However, this is not true for a microsecond high-voltage pulse. For example, consider the conditions shown in figure 9.1 under which breakdown in water occurs: $E = 80\ {\rm MV\ m^{-1}}$, $t_{\rm delay} = 4\ \mu$s. Assuming that the voltage pulse on the electrode has a

**Table 9.1.** Experimental conditions under which a nanosecond discharge in water was observed in the studies cited in the column headings.

| | [7] | [12] | [14] |
|---|---|---|---|
| $r_{\rm el}$[μm] | 50 | 33 | 35 |
| $d$[mm] | 4 | 4 | 1.5 |
| U[kV] | 25 | 30 | 24 |
| $\tau_0$[ns] | 1–3 | 1–3 | 1–3 |
| $E$ [MVm$^{-1}$] | 200 | 331 | 290 |
| $P_-$[MPa] | –14 | –39 | –31 |

rectangular shape with a duration of $\tau_0 = t_{delay}$, an estimate for the released energy in regular tap water (conductivity of $\Sigma_{tap} = 0.05\ \mathrm{S\ m^{-1}}$) is $W_{T,tap} \approx 1.3 \cdot 10^9\ \mathrm{J\ m^{-3}}$. So at microsecond high-voltage pulses, Joule heating becomes sufficient for the formation of gas bubbles by local evaporation and hydrolysis in water.

## 9.4 The cavitation discharge model and analysis of experimental data

In previous chapters, we considered the electrostrictive mechanism of cavitation formation and conditions of its initiation, and presented a theory for the nucleation and growth of cavitation nanovoids. We use these results to analyze the experimental data of [7, 12, 14] in the points outlined below.

1. In all experiments, the negative pressure $|P_{-}|$ exceeds 10 MPa, for which cavitation does occur in water according to the literature (table 9.1).
2. Electric breakdown develops much sooner than the fluid leaks to the electrode and negative pressure compensation occurs. In fact, by substituting the values of the electrode's tip radius of curvature from table 9.1 and the speed of sound in water $c_s \approx 1\ 500\ \mathrm{m\ s^{-1}}$ into formula (1.40), we obtain that the time by which the negative pressure is compensated exceeds 30 ns in all the experiments. Hence, the hydrostatic pressure changes can be neglected.
3. The time $t_{ion}$, during which a growing nanovoid reaches a size large enough for electron to acquire energy that exceeds the ionization potential of the molecules, is much shorter than the rise time of the high-voltage pulse. To estimate $t_{ion}$, we can use the asymptotic formula (6.18) for the rate of expansion of cavitation pores (chapter 6):

$$e\Delta\phi = 2eE_{in}R(t) \sim 2eE_{in} \cdot \sqrt{\frac{2|P_{-}|}{3\rho}}\, t_{ion} \geqslant I_i. \tag{9.9}$$

   Here, $\Delta\phi$ is the potential difference between the nanopore's poles and $R(t)$ is the radius of the nanopore, $E_{in} \approx 3/2 \cdot E$ is the electric field inside the spherical pore (equation (4.53)), $E$ is the 'external' field outside the pore (the field in table 9.1), $\rho = 10^3\ \mathrm{kg\ m^{-3}}$ is the density of water, $P_{-}$ is the negative pressure from table 9.1, and $I_i = 12.6$ eV is the ionization potential of water molecules. Given these values (9.9) yields $t_{ion} \approx 0.1$ ns. Therefore, in accordance with the cavitation model, the necessary conditions for breakdown in water are fulfilled in all experiments listed in table 9.1.
4. The fading of the discharge (dark phase) in figures 9.2 and 9.5(b) (6–14 ns), is associated with the leaking of fluid to the electrode, which causes the absolute value of the negative pressure $|P_{-}|$ to decrease below the level at which cavitation voids can be generated. The subsequent re-ignition of the discharge, shown in figures 9.2 and 9.5(c) (16–18 ns), is associated with the fluid moving away from the electrode when the voltage drops on it, which reduces the electrostriction (negative) component of the total pressure. Figure 5.13 shows computed density profiles in the vicinity of an electrode in water at different time moments at the voltage amplitude 20 kV [17]. As we show in

chapter 5, if the voltage pulse is long enough, the hydrostatic pressure next to the electrode can increase as a result of influx of fluid and can partially or completely compensate the negative electrostrictive pressure. Then, at the end of the pulse when the voltage reduces, the fluid flows from the electrode. Areas with density lower than in the unperturbed fluid form due to its inertia. The total pressure again becomes negative and can exceed the cavitation limit. So the conditions for electron ionization avalanche, and therefore for discharge initiation, can arise again at the end of the pulse in the experiments (see figure 9.2). If the voltage pulse falls faster than the outflux of fluid away from the electrode, then no secondary discharge can initiate (see figure 9.3).

## 9.5 A qualitative picture of nanosecond breakdown in liquids

The scenario for pulsed nanosecond breakdown in liquid dielectrics should occur in the following manner. In the beginning, the strong inhomogeneous electric field at the needle electrode creates a region saturated with micropores in the liquid. The electric field accelerates the primary electrons in the pores to energies that exceed the potential of ionization of water molecules, forming microstreamers. After neutralization of electrons at the electrode, positive ions form a virtual needle electrode in the liquid, which fulfils the electrostrictive conditions for the appearance of the next set of cavities. This streamer propagation continues until the voltage drop on the moving streamer reduces below the order of the initial voltage on the needle electrode. In the process of propagation, the head of a streamer narrows (in accordance with observations from [7]) so the electrostrictive conditions for breakdown recur in the vicinity of the streamer head. This breakdown mechanism in liquid dielectrics is valid only when the rise time of the voltage pulse is much shorter than the time of equalizing the total pressure in the vicinity of the electrode or in the vicinity of the streamer head.

Thus, as a result of breakdown development, the nanosecond discharge enters the streamer mode (figure 9.8).

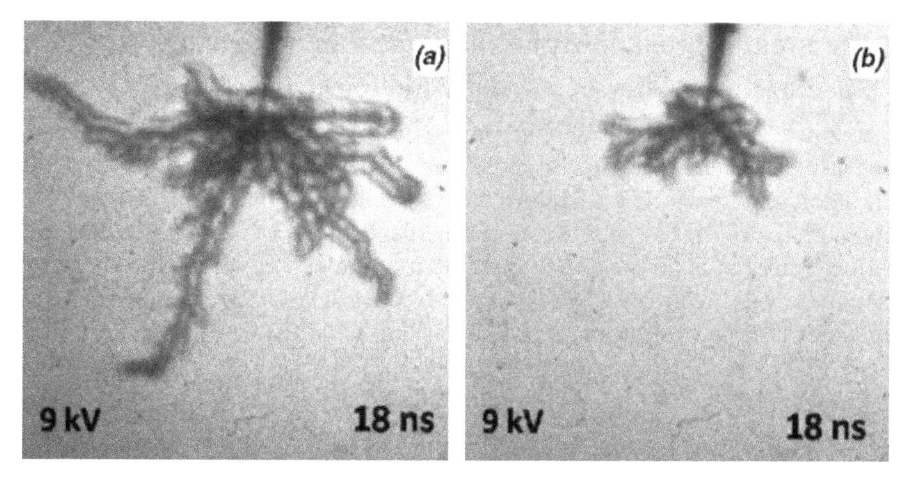


**Figure 9.8.** Schlieren images of 'tree-like' streamer discharges in water (a) and ethanol (b) at an electrode of radius $\approx 1$ μm, and under a positive voltage pulse of rise time 5 ns and amplitude 9 kV. Reproduced from [18].

The transfer of the high-voltage potential into the streamer head with a micron scale radius leads to a sharply inhomogeneous field in its vicinity. As a result, ponderomotive forces arise that lead to intense cavitation nucleation, and therefore to the possibility of ionization electron multiplication and further propagation of the streamer. The region of space charge in the streamer head is not only the source of a strong electric field, but also leads to Coulomb repulsion, which increases the local values of the negative pressure and thereby accelerates the process of cavitation nucleation and growth of cavitation nanopores.

Initiation of the cavitation breakdown mechanism is independent of the voltage pulse polarity. Yet, it depends on the rise time of the voltage pulse, the permittivity of the liquid, and the gradient of the square of the electric field. The development of the discharge, particularly streamers, does however depend on the polarity. At positive polarity, the electrons flow out from the head of a developing streamer. The remaining positive space charge causes a sharp increase in the field in front of the head and in the development of cavitation and ionization multiplication near it. At negative polarity, the streamer head has an excess of electrons, which increases the effective radius of the head due to high electron mobility. This reduces the field near the head while the potential of the head stays the same. As a result, the onset of cavitation and associated ionization multiplication are suppressed. This requires a greater magnitude of the applied voltage for streamer development which, in particular, explains the results of experiments on nanosecond breakdown in water shown in figure 9.6.

Now we will say a few words about microsecond breakdown in liquids. All experimental studies of this phenomenon (e.g. [1]) note that the first stage of the breakdown begins with a deformation of the existing bubble in the electrode or the generation of such a bubble. In the second stage, the streamers rapidly spread out from the top of the bubble, as shown in figure 9.1(c). The generation and propagation of streamer channels in microsecond discharges appear to have the same nature as the streamers developing under conditions of nanosecond breakdown.

## 9.6 The area of breakdown initiation at a nanosecond voltage pulse in the vicinity of a needle-like electrode [26]

Earlier, in chapter 6, we consider the problem of nanopore expansion in a linearly increasing electric field. We now estimate the region in the vicinity of the electrode in which the nanopores grow enough for discharge ignition (9.9).

Knowing the spatial distribution and time dependence of the electric field during the voltage pulse, it is possible to find the area in the vicinity of the electrode where the nanopore will grow to the size at which electrons can acquire enough energy to ionize water molecules. For simplicity, we will assume that

$$E(r, t) = E_0 \cdot \frac{r_{\mathrm{el}}^2}{r^2} \cdot \frac{t}{\tau_0}, \tag{9.10}$$

where $r_{\mathrm{el}}$ is the curvature radius of the electrode's tip.

The corresponding negative pressure distribution is (see chapter 6)

$$P_{-}(r,\,t) \approx k_{\mathrm{E}} P_{\mathrm{E}} \frac{r_{\mathrm{el}}^4}{r^4} \cdot \frac{t^2}{\tau_0^2}. \tag{9.11}$$

All the following calculations are based on equation (6.17) with $\delta_{\mathrm{b}} = 2.4$ nm, $R_0 = R_{\mathrm{cr}} = 2.5$ nm, $\sigma_{0\mathrm{s}} = 0.072\ \mathrm{N\,m^{-1}}$, $P_0 = 40$ MPa, $r_{\mathrm{el}} = 35\ \mu\mathrm{m}$, $\tau_0 = 3$ ns, $0 \leqslant t \leqslant \tau_0$, and $k_{\mathrm{E}} = 0.346$. The chosen values of $R_{\mathrm{cr}}$, $\delta_{\mathrm{b}}$, and $\sigma_{0\mathrm{s}}$ correspond to the critical negative pressure of −30 MPa.

Figures 9.9 and 9.10 show the dependence of $|P_{\mathrm{E}}(r,\,\tau_0)|$, the pore radius at $R$, and the kinetic energy $\Delta\phi = 1.5E_{\mathrm{in}}R$ that an electron acquires in the growing pore on the distance from the electrode at $t = \tau_0$.

It follows from figure 9.10(b), that at the time moment $t = \tau_0 = 3$ ns, the area where the nanopores reach a size at which ionization begins is of the order of $\approx 2\ \mu\mathrm{m}$, and the size at which the negative pressure exceeds the cavitation threshold is much greater at $\approx 14\ \mu\mathrm{m}$.

In general, three characteristic regions can be distinguished near the electrode. Figure 9.11 shows a schematic picture of the regions in the vicinity of an electrode where cavitation micropores appear and develop. In region 1, where the electric field gradient is greatest, the occurring cavitation nanopores have enough time during the nanosecond voltage pulse to grow to a size at which electrons can gain enough energy to excite and ionize molecules of the liquid on the pore wall. In region 2, the electrostrictive negative pressure reaches values at which cavitation development becomes possible (which can be registered by the optical methods), but nanovoids

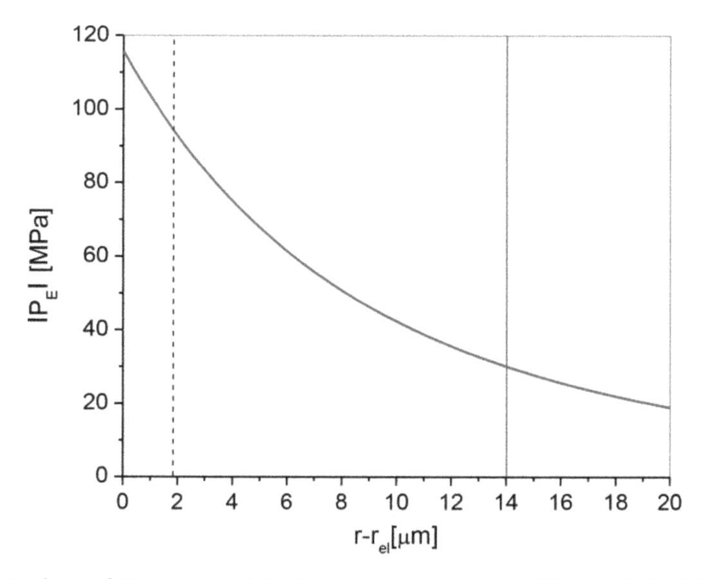


**Figure 9.9.** Dependence of $P_{\mathrm{E}}$ on $r - r_{\mathrm{el}}$ at the time moment $t = \tau_0 = 3$ ns. The area to the left of the dashed vertical line shows where the pores of the initial size of 2.5 nm are expanding and the area to the right where they are collapsing. The area where the electrostriction tension (negative pressure) exceeds the cavitation threshold $|P_{\mathrm{E}}| > 30$ MPa is to the left of the solid vertical line.

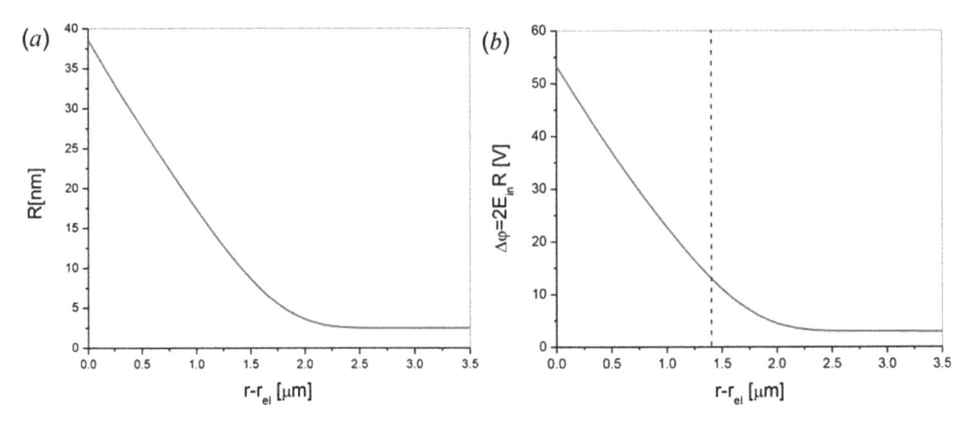


**Figure 9.10.** Dependence of $R$ (a) and $\Delta\phi$ (b) on $r - r_{el}$ at the time moment $t = \tau_0 = 3$ ns. The dashed vertical line in plot (b) corresponds to $e\Delta\phi = I_i = 12.6$ eV.

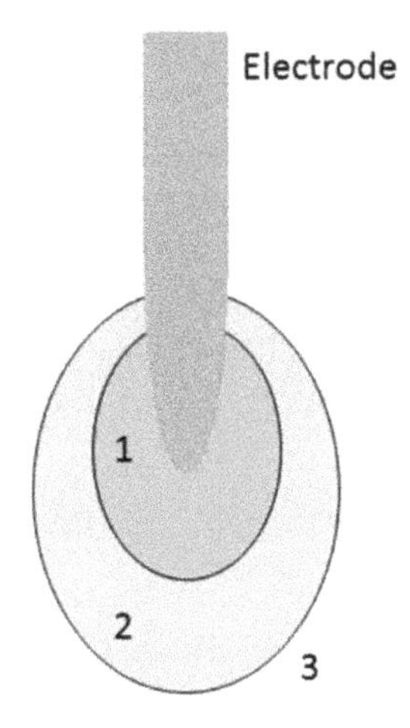


**Figure 9.11.** Schematic picture of the three characteristic areas in the vicinity of a needle electrode where cavitation nanopores appear and develop. Reproduced with permission from [24]. Copyright 2015 AIP Publishing LLC.

appearing during the voltage pulse do not have enough time to grow to the size at which the potential difference across their borders becomes sufficient for ionization or excitation of water molecules. In region 3, the development of cavitation is impossible, because the spontaneously occurring nanovoids do not grow since the value of the electrostrictive negative pressure is relatively small and cannot compete with the forces of surface tension.

## 9.7 The problem of primary electrons

We showed that, in electric fields typical for nanosecond breakdown in liquid, multiplication of electrons can occur by collisionless energy gain in the resulting cavitation cavities. But the question remains of where the primary electrons (the first ones that seed the multiplication process) come from. In fact, this issue is common to all discharge physics in liquids and is not directly related to the subject of this book. Therefore, we will only discuss it qualitatively.

Possible sources of the primary electrons include cosmic background radiation and field ionization of water molecules and dissolved $OH^-$ ions. At atmospheric pressure and a temperature of 25 °C, the density of water molecules is $n = 3.3 \cdot 10^{28}\ \mathrm{m}^{-3}$ and the equilibrium density of ions in distilled water is $n_{OH^-} = n_{H^+} = 6 \cdot 10^{19}\ \mathrm{m}^{-3}$ [25].

We estimate the minimum number of electrons that should reside in a cubic micron so that the development of discharge is possible. According to estimates (see figure 2.11), the number of nanopores in a $\mu\mathrm{m}^{-3}$ is on the order of $N_b \approx 2 \cdot 10^4$. Assuming the radius of the nanopores is ~10 nm, we find that their total volume makes up about 8% of the volume of water. If we assume that the electrons are randomly distributed in the volume occupied by the nanopores and water, then the presence of 13 primary electrons is sufficient to make sure that at least one resides in the volume occupied by a nanopore and serves as the seed for breakdown development.

Regarding natural background radiation, the probability that it is responsible for ionizing the primary electrons is extremely small. For example, in air the equilibrium density of ions (positive and negative, formed by electron attachment) that are initiated by cosmic radiation is quite small: of the order of $10^9\ \mathrm{m}^{-3}$ [26]. It is difficult to expect that background radiation will lead to a significantly greater degree of ionization (orders of magnitude more ions) in liquids.

Estimation of the probability of field ionization of water molecules as a result of electron tunneling from hydrogen-like atoms [27], with the water ionization potential of $I_i = 12.6$ eV, shows that it is insignificant in fields weaker than $7.5 \cdot 10^9\ \mathrm{V\ m}^{-1}$. To estimate the probability of electron detachment from negative $OH^-$ ions, we use the formula for tunneling detachment of electrons [27, 28]:

$$w \approx \pi A^2 E \sqrt{\frac{e}{2 I_n m}} \exp\left(-\frac{4}{3}\frac{\sqrt{2me}}{\hbar}\frac{I_n^{3/2}}{E}\right), \tag{9.12}$$

in which $E$ is the electric field, $\hbar$ is Planck's constant, $e$ and $m$ are the charge and mass of the electron, and $I_n$ is the affinity energy of an electron in a negative ion (in eV). The coefficient $A \sim 1$ depends on the shape of the potential well. The number of emitted electrons in a cubic micron under the field $E$ in the area of the breakdown during time $\Delta t$ is

$$N_e \sim w(I_n, E) N_{OH^-} \Delta t, \tag{9.13}$$

where $N_{OH^-} = 60$ is the number of negative ions in a $\mu\mathrm{m}^{-3}$. Given that the energy of an electron in the $OH^-$ ion is $I_n = 1.85$ eV [29], already at $E = 1.15 \cdot 10^9\ \mathrm{V\ m}^{-1}$, the number of electrons formed by field detachment in a cubic micron in time $\Delta t = 1$ ns is $N_e > 15$. So at least one of these ions will reside in the vicinity of a forming pore and when the field is $E = 1.25 \cdot 10^9\ \mathrm{V\ m}^{-1}$, virtually all $OH^-$ ions emit electrons in $\Delta t \sim 1$ ns.

In this estimation, we used the equilibrium number of $OH^-$ ions in the absence of an electric field. This estimate can be considered as the lower bound, because in the presence of a strong field the degree of water molecule dissociation increases significantly and the amount of $OH^-$ ions may increase by several orders of magnitude [30].

In the experiments of [7, 12–18] (see table 9.1), discharge occurred at fields of $E \approx 0.3 \cdot 10^9$ V m$^{-1}$, which is more than three times smaller than our estimate based on field emission of electrons from $OH^-$ ions. Thus, the question of the source of primary electrons in nanosecond and microsecond discharges in water requires further research.

## References


[1] Ushakov V Y, Klimkin V F and Korobeynikov S M 2005 *Breakdown in Liquids at Impulse Voltage* (Tomsk: NTL) (in Russian)
Ushakov V Y, Klimkin V F and Korobeynikov S M 2007 *Impulse Breakdown of Liquids (Power Systems)* (Berlin: Springer)

[2] Belevtsev A A 2015 Electronic transport coefficients and electric breakdown in condensed helium *High Temp.* **53**779

[3] Naidis G V 2015 Modelling of streamer propagation in hydrocarbon liquids in point-plane gaps *J. Phys. D: Appl. Phys.* **48**195203

[4] Ushakov V Y 1975 *Pulsed Electric Breakdown of Liquids* (Tomsk: Tomsk State University) (in Russian)

[5] Efanov V and Efanov M 2008 New possibilities of picosecond and nanosecond FID technology for medical applications *IEEE Int. Conf. Power Modulators and High Voltage* p 334

[6] Efanov V and Efanov M 2008 Gigawatt all solid state FID pulsers with nanosecond pulse duration *IEEE Int. Conf. Power Modulators and High Voltage* p 381

[7] Starikovskiy A, Yang Y, Cho Y I and Fridman A 2011 Non-equilibrium plasma in liquid water: dynamics of generation and quenching *Plasma Sources Sci. Technol.* **20**024003

[8] Shneider M N, Pekker M and Fridman A 2012 Theoretical study of the initial stage of sub-nanosecond pulsed breakdown in liquid dielectrics *IEEE Trans. Dielectr. Electr. Insul.* **19**1597

[9] Shneider M N and Pekker M 2013 Dielectric fluid in inhomogeneous pulsed electric field *Phys. Rev.* E **87**043004

[10] Alkhimov A P, Vorobev V V, Klimkin V F, Ponomarev A G and Soloukhin R I 1970 Development of electric discharge in water *Dokl. Akad. Nauk SSSR* **194** 1052 (in Russian)

[11] Abramyan E A, Kornilov V A, Lagunov V M, Ponomarev A G and Soloukhin R I 1971 Megavolt energy densifier *Dokl. Akad. Nauk SSSR* **201** 56 (in Russian)

[12] Marinov I L, Guaitella O, Rousseau A and Starikovskaia S M 2011 Successive nanosecond discharges in water *IEEE Trans. Plasma Sci.* **39**2672

[13] Dobrynin D, Seepersad Y, Pekker M, Shneider M, Friedman G and Fridman A 2013 Non-equilibrium nanosecond-pulsed plasma generation in the liquid phase (water, PDMS) without bubbles: fast imaging, spectroscopy and leader-type model *J. Phys. D: Appl. Phys.* **46**105201

[14] Seepersad Y, Pekker M, Shneider M N, Fridman A and Dobrynin D 2013 Investigation of positive and negative modes of nanosecond pulsed discharge in water and electrostriction model of initiation *J. Phys. D: Appl. Phys.* **46**355201
[15] Seepersad Y, Fridman A and Dobrynin D 2015 Anode initiated impulse breakdown in water: the dependence on pulse rise time for nanosecond and sub-nanosecond pulses and initiation mechanism based on electrostriction *J. Phys. D: Appl. Phys.* **48**424012
[16] Marinov I, Starikovskaia S and Rousseau A 2014 Dynamics of plasma evolution in a nanosecond underwater discharge *J. Phys. D: Appl. Phys.* **47**224017
[17] Pekker M, Seepersad Y, Shneider M N, Fridman A and Dobrynin D 2014 Initiation stage of nanosecond breakdown in liquid *J. Phys. D: Appl. Phys.* **47**025502
[18] Marinov I, Guaitella O, Rousseau A and Starikovskaia S M 2013 Cavitation in the vicinity of the high-voltage electrode as a key step of nanosecond breakdown in liquids *Plasma Sources Sci. Technol.* **22**042001
[19] Raiser Yu P 1991 *Gas Discharge Physics* (Berlin: Springer)
[20] Martin J C 1970 *Nanosecond Pulse Techniques* (*Advances in Pulsed Power Technology* vol 3) (Berlin: Springer) pp 35–74
[21] Bazelyan E M and Raizer Yu P 1997 *Spark Discharge* (Boca Raton, FL: CRC)
[22] Radiometer Analytical 2004 *Conductivity Theory and Practice* (Villeurbanne: Radiometer Analytical SAS)
[23] Haynes W M 2003 *CRC Handbook of Chemistry and Physics* 93th edn (Boca Raton, FL: CRC Press)
[24] Shneider M N and Pekker M 2015 Pre-breakdown processes in a dielectric fluid in inhomogeneous pulsed electric fields *J. Appl. Phys.* **117**224902
[25] Eisenberg D and Kauzmann W 1969 *The Structure and Properties of Water* (Oxford: Oxford University Press)
[26] Rühling F, Heilbronner F and Ortéga P 2003 Discharge inception under impulse voltage: influence of lab air ion density and resulting charge-voltage relation *13th Int. Symp. on High Voltage Engineering (Rotterdam)*
[27] Landau L D and Lifshitz E M 1991 *Quantum Mechanics: Non-relativistic Theory* 3rd edn (Oxford: Pergamon)
[28] Demkov Y N and Drukarev G F 1965 Decay and polarizability of negative ions in an electric field *Sov. Phys.—JETP* **20** 614
[29] Goldfarb F, Drag C, Chaibi W, Kröger S, Blondel C and Delsart C 2005 Photodetachment microscopy of the P, Q, and R branches of the OH($v = 0$) to OH($v = 0$) detachment threshold *J. Chem. Phys.* **122**014308
[30] Saitta A M, Saija F and Giaquinta P V 2012 *Ab initio* molecular dynamics study of dissociation of water under an electric field *Phys. Rev. Lett.* **108**207801

Lightning Source UK Ltd.
Milton Keynes UK
UKHW05n1331190718
325974UK00003B/62/P